QuickStart Math

Bringing Skills Into Focus

Acknowledgments

Executive Editor: Linda Bullock, Ph.D.
Editor: Joshua Fisher
Senior Designer: Deborah Diver
Editorial Development: MATHQueue, Inc.
Design and Production: Marc Publisher Services
Cover Design: Marc Publisher Services

ISBN-10: 1-60161-138-2
ISBN-13: 978-1-60161-138-3

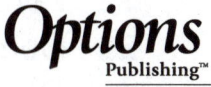

Printed in the U.S.A.
15 14 13 12 11 10 9 8 7 6 5 4 3 2 1

Options Publishing™
A Haights Cross Communications Company

Options Publishing
P.O. Box 1749
Merrimack, NH 03054-1749
Toll Free 800-782-7300
Toll Free Fax 866-424-4056
www.optionspublishing.com

Contents

Activities

Tools

Complete this page with your teacher.

▶ **Find the least common multiple, or LCM, of 12 and 18.**

A. List multiples.

Multiples of 12: _____, _____, _____, _____, _____, …

Multiples of 18: _____, _____, _____, _____, _____, …

The LCM of 12 and 18 is _____.

B. Use prime factorization.

12 = _____ × _____ × _____

18 = _____ × _____ × _____

The LCM is _____ × _____ × _____ × _____ = _____.

▶ **Find each LCM.**

C. 15 and 20

LCM = _____

D. 5 and 14

LCM = _____

E. 9 and 12

LCM = _____

F. 6 and 25

LCM = _____

G. 8 and 10

LCM = _____

H. _____ and _____

LCM = _____

Practice

▶ **Circle the least common multiple of the numbers in each pair.**

1. 6 and 8

    A. 12   **C.** 24
    B. 18   **D.** 48

2. 10 and 12

    A. 60   **C.** 24
    B. 40   **D.** 20

3. 3 and 14

    A. 28   **C.** 56
    B. 42   **D.** 64

4. 8 and 18

    A. 16   **C.** 72
    B. 36   **D.** 96

5. 5 and 9

    A. 27   **C.** 42
    B. 35   **D.** 45

6. 5 and 17

    A. 95   **C.** 80
    B. 85   **D.** 34

7. 10 and 25

    A. 50   **C.** 25
    B. 40   **D.** 20

8. 4 and 9

    A. 18   **C.** 36
    B. 24   **D.** 40

9. 3 and 17

    A. 63   **C.** 34
    B. 51   **D.** 21

10. 4 and 5

    A. 8   **C.** 15
    B. 10   **D.** 20

11. 16 and 24

    A. 8   **C.** 40
    B. 32   **D.** 48

12. 45 and 60

    A. 180   **C.** 90
    B. 105   **D.** 15

13. 30 and 9

    A. 270   **C.** 60
    B. 90   **D.** 3

14. 36 and 16

    A. 4   **C.** 72
    B. 52   **D.** 144

Complete this page with your teacher.

▶ **Make equivalent fractions. Write the missing numerator or denominator.**

A.
$$\frac{4}{5} = \frac{}{20}$$
×4 (top), ×4 (bottom)

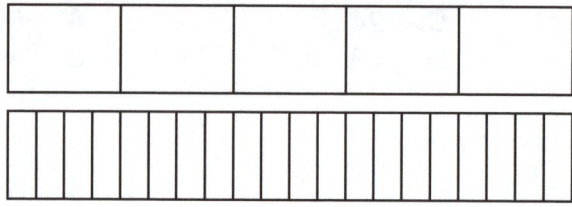

B.
$$\frac{9}{12} = \frac{}{4}$$
÷3 (top), ÷3 (bottom)

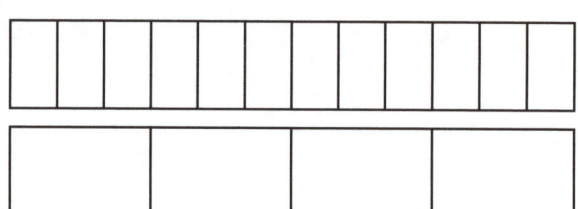

C. $\dfrac{}{16} = \dfrac{3}{8}$ D. $\dfrac{1}{3} = \dfrac{5}{}$

E. $\dfrac{}{24} = \dfrac{9}{6}$ F. $\dfrac{3}{8} = \dfrac{42}{}$

G. $\dfrac{3}{5} = \dfrac{27}{}$ H. $\dfrac{}{51} = \dfrac{24}{17}$

I. $\dfrac{}{14} = \dfrac{1}{2}$ J. $\dfrac{\square}{\square} = \dfrac{\square}{\square}$

Practice

▶ **Make equivalent fractions. Write the missing numerator or denominator.**

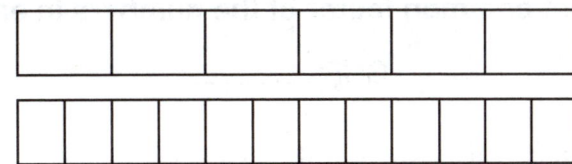

1. $\dfrac{5}{6} = \dfrac{}{12}$

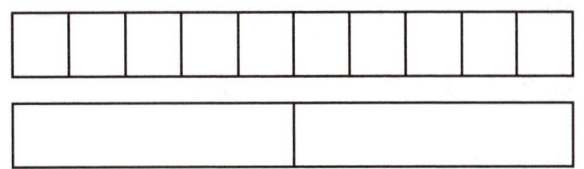

2. $\dfrac{5}{10} = \dfrac{}{2}$

3. $\dfrac{}{8} = \dfrac{20}{32}$

4. $\dfrac{}{6} = \dfrac{15}{18}$

5. $\dfrac{4}{20} = \dfrac{1}{}$

6. $\dfrac{18}{4} = \dfrac{9}{}$

7. $\dfrac{1}{} = \dfrac{4}{16}$

8. $\dfrac{4}{7} = \dfrac{}{21}$

9. $\dfrac{}{18} = \dfrac{3}{6}$

10. $\dfrac{16}{} = \dfrac{2}{3}$

11. $\dfrac{30}{} = \dfrac{1}{2}$

12. $\dfrac{16}{20} = \dfrac{4}{}$

13. $\dfrac{18}{} = \dfrac{2}{5}$

14. $\dfrac{8}{9} = \dfrac{}{45}$

15. $\dfrac{}{45} = \dfrac{2}{3}$

16. $\dfrac{42}{} = \dfrac{6}{8}$

17. $\dfrac{6}{15} = \dfrac{}{5}$

Complete this page with your teacher.

▶ **Find the greatest common factor of the numbers in each set.**

A. 16 and 20   GCF: _____

$1 \times 16 = 16, 2 \times 8 = 16, 4 \times 4 = 16$

Factors of 16:   _____, _____, _____, _____, _____

$1 \times 20 = 20, 2 \times 10 = 20, 4 \times 5 = 20$

Factors of 20:   _____, _____, _____, _____, _____, _____

B. 18 and 45   GCF: _____

Factors of 18:   _____

Factors of 45:   _____

C. 32, 48, and 64   GCF: _____

Factors of 32:   _____

Factors of 48:   _____

Factors of 64:   _____

D. _____ and _____   GCF: _____

Factors of _____: _____

Factors of _____: _____

Practice

▶ **Match the numbers in each set with their GCF, or greatest common factor. Write the letter of the answer on the line.**

_____	1. 30 and 45	**A.**	12
_____	2. 56 and 63	**B.**	14
_____	3. 36 and 81	**C.**	4
_____	4. 24 and 60	**D.**	15
_____	5. 18 and 54	**E.**	8
_____	6. 42 and 56	**F.**	7
_____	7. 24, 42, and 72	**G.**	9
_____	8. 26, 52, and 65	**H.**	3
_____	9. 25, 30, and 45	**I.**	2
_____	10. 18, 20, and 30	**J.**	5
_____	11. 90 and 45	**K.**	45
_____	12. 76 and 57	**L.**	19
_____	13. 16 and 24	**M.**	13
_____	14. 51 and 60	**N.**	18
_____	15. 36 and 16	**O.**	6

Complete this page with your teacher.

▶ **Write each fraction in simplest form.**

A. $\dfrac{15}{42} = \dfrac{15 \div \Box}{42 \div \Box} = \dfrac{\Box}{\Box}$

Factors of 15: _____

Factors of 42: _____

GCF: _____

B. $\dfrac{25}{30} = \dfrac{25 \div \Box}{30 \div \Box} = \dfrac{\Box}{\Box}$

Factors of 25: _____

Factors of 30: _____

GCF: _____

C. $\dfrac{18}{24} = \dfrac{\Box}{\Box}$

D. $\dfrac{9}{24} = \dfrac{\Box}{\Box}$

E. $\dfrac{16}{36} = \dfrac{\Box}{\Box}$

F. $\dfrac{8}{10} = \dfrac{\Box}{\Box}$

G. $\dfrac{16}{10} = \dfrac{\Box}{\Box}$

H. $\dfrac{6}{15} = \dfrac{\Box}{\Box}$

I. $\dfrac{20}{30} = \dfrac{\Box}{\Box}$

J. $\dfrac{4}{20} = \dfrac{\Box}{\Box}$

K. $\dfrac{\Box}{\Box} = \dfrac{\Box}{\Box}$

Practice

▶ **Write each fraction in simplest form.**

1. $\frac{18}{54}$ = _____

2. $\frac{10}{45}$ = _____

3. $\frac{15}{18}$ = _____

4. $\frac{24}{42}$ = _____

5. $\frac{12}{32}$ = _____

6. $\frac{9}{36}$ = _____

7. $\frac{14}{35}$ = _____

8. $\frac{14}{54}$ = _____

9. $\frac{36}{48}$ = _____

10. $\frac{64}{72}$ = _____

11. $\frac{20}{32}$ = _____

12. $\frac{6}{8}$ = _____

13. $\frac{16}{20}$ = _____

14. $\frac{30}{45}$ = _____

15. $\frac{24}{32}$ = _____

16. $\frac{8}{4}$ = _____

▶ **Circle TRUE or FALSE for each sentence.**

17. $\frac{9}{10}$ is in simplest form.   TRUE   FALSE

18. $\frac{36}{45}$ simplifies to $\frac{3}{5}$.   TRUE   FALSE

19. $\frac{2}{26}$ is in simplest form.   TRUE   FALSE

20. $\frac{16}{28}$ simplifies to $\frac{4}{7}$.   TRUE   FALSE

Complete this page with your teacher.

▶ **Find each sum. Express each answer in simplest form.**

A. $\dfrac{6}{15} \longrightarrow \dfrac{6}{15}$ B. $\dfrac{3}{8} \longrightarrow \underline{\quad}$ C. $\dfrac{2}{3} \longrightarrow \underline{\quad}$

$+ \dfrac{2}{5} \longrightarrow \dfrac{6}{15}$ $+ \dfrac{1}{4} \longrightarrow \underline{\quad}$ $+ \dfrac{3}{12} \longrightarrow \underline{\quad}$

$\underline{\quad}\quad \dfrac{12}{15} = \dfrac{4}{5}$ $\underline{\quad}\quad \underline{\quad}$ $\underline{\quad}\quad \underline{\quad}$

D. $\dfrac{6}{10} \longrightarrow \underline{\quad}$ E. $\dfrac{6}{8} \longrightarrow \underline{\quad}$ F. $\dfrac{10}{2} \longrightarrow \underline{\quad}$

$+ \dfrac{1}{4} \longrightarrow \underline{\quad}$ $+ \dfrac{3}{5} \longrightarrow \underline{\quad}$ $+ \dfrac{2}{3} \longrightarrow \underline{\quad}$

$\underline{\quad}\quad \underline{\quad}$ $\underline{\quad}\quad \underline{\quad}$ $\underline{\quad}\quad \underline{\quad}$

G. $\dfrac{1}{6} \longrightarrow \underline{\quad}$ H.

$+ \dfrac{3}{8} \longrightarrow \underline{\quad}$

$\underline{\quad}\quad \underline{\quad}$

Practice

▶ **Find each sum. Express each answer in simplest form.**

1. $\dfrac{1}{6}$
    $+\ \dfrac{1}{3}$
    ___

2. $\dfrac{2}{3}$
    $+\ \dfrac{1}{5}$
    ___

3. $\dfrac{11}{18}$
    $+\ \dfrac{4}{12}$
    ___

4. $\dfrac{4}{9}$
    $+\ \dfrac{1}{3}$
    ___

5. $\dfrac{1}{3}$
    $+\ \dfrac{2}{8}$
    ___

6. $\dfrac{1}{2}$
    $+\ \dfrac{2}{5}$
    ___

7. $\dfrac{1}{8}$
    $+\ \dfrac{3}{6}$
    ___

8. $\dfrac{3}{4}$
    $+\ \dfrac{2}{16}$
    ___

9. $\dfrac{1}{4}$
    $+\ \dfrac{2}{3}$
    ___

10. $\dfrac{4}{8}$
    $+\ \dfrac{3}{12}$
    ___

11. $\dfrac{4}{10}$
    $+\ \dfrac{2}{8}$
    ___

12. $\dfrac{1}{2}$
    $+\ \dfrac{1}{6}$
    ___

Complete this page with your teacher.

▶ **Find each difference. Write each answer in simplest form.**

A. $\dfrac{7}{10}$ ⟶ $\dfrac{7}{10}$

$-\ \dfrac{2}{5}$ ⟶ $\dfrac{4}{10}$
_____

$\dfrac{3}{10}$

B. $\dfrac{2}{3}$ ⟶ _____

$-\ \dfrac{1}{6}$ ⟶ _____
_____

_____

C. $\dfrac{3}{4}$ ⟶ _____

$-\ \dfrac{1}{3}$ ⟶ _____
_____

_____

D. $\dfrac{6}{8}$ ⟶ _____

$-\ \dfrac{2}{4}$ ⟶ _____
_____

_____

E. $\dfrac{2}{3}$ ⟶ _____

$-\ \dfrac{1}{12}$ ⟶ _____
_____

_____

F. $\dfrac{11}{5}$ ⟶ _____

$-\ \dfrac{1}{3}$ ⟶ _____
_____

_____

G. $\dfrac{4}{5}$ ⟶ _____

$-\ \dfrac{4}{6}$ ⟶ _____
_____

_____

H.
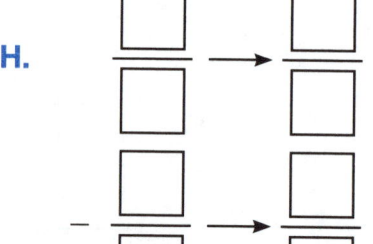
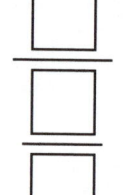

Practice

▶ **Find each difference. Write each answer in simplest form.**

1. $\dfrac{1}{2}$
    $-\ \dfrac{1}{3}$

2. $\dfrac{2}{3}$
    $-\ \dfrac{2}{5}$

3. $\dfrac{5}{6}$
    $-\ \dfrac{7}{12}$

4. $\dfrac{5}{9}$
    $-\ \dfrac{1}{3}$

5. $\dfrac{5}{6}$
    $-\ \dfrac{4}{8}$

6. $\dfrac{4}{9}$
    $-\ \dfrac{1}{6}$

7. $\dfrac{15}{16}$
    $-\ \dfrac{3}{4}$

8. $\dfrac{3}{4}$
    $-\ \dfrac{3}{9}$

9. $\dfrac{5}{8}$
    $-\ \dfrac{1}{2}$

10. $\dfrac{7}{10}$
    $-\ \dfrac{1}{5}$

11. $\dfrac{2}{4}$
    $-\ \dfrac{2}{5}$

12. $\dfrac{1}{2}$
    $-\ \dfrac{2}{7}$

Complete this page with your teacher.

▶ **Find each sum. Write each answer in simplest form.**

A. $4\frac{1}{2}$ ⟶ 4 ——

　　+ $9\frac{3}{8}$ ⟶ 9 ——

B. $12\frac{5}{6}$ ⟶ 12 ——

　　+ $8\frac{1}{3}$ ⟶ 8 ——

C. $31\frac{2}{9}$ ⟶ 31 ——

　　+$17\frac{1}{3}$ ⟶ 17 ——

D. $9\frac{4}{5}$ ⟶ 9 ——

　　+ $6\frac{3}{4}$ ⟶ 6 ——

E. $12\frac{3}{8}$ ⟶ 12 ——

　　+ $3\frac{1}{4}$ ⟶ 3 ——

F.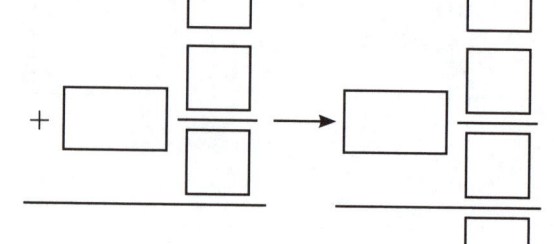

Practice

▶ **Find each sum. Write each answer in simplest form. Then shade the matching answer in the table.**

1. $8\frac{2}{8}$
   $+\ 4\frac{1}{4}$

2. $19\frac{5}{6}$
   $+\ 17\frac{1}{2}$

3. $36\frac{1}{2}$
   $+\ 9\frac{1}{7}$

4. $34\frac{5}{9}$
   $+\ 25\frac{2}{3}$

5. $52\frac{2}{3}$
   $+\ 38\frac{3}{10}$

6. $65\frac{3}{4}$
   $+\ 16\frac{2}{3}$

7. $17\frac{3}{4}$
   $+\ 4\frac{4}{5}$

8. $45\frac{5}{12}$
   $+\ 23\frac{2}{6}$

9. $19\frac{1}{2}$
   $+\ 56\frac{1}{3}$

$82\frac{5}{12}$	$37\frac{1}{3}$	$75\frac{5}{6}$	$12\frac{1}{2}$	$60\frac{2}{9}$	$90\frac{29}{30}$	$45\frac{9}{14}$	$22\frac{11}{20}$	$68\frac{3}{4}$

Complete this page with your teacher.

▶ **Find each difference. Write each answer in simplest form.**

A. $8\frac{1}{2} \longrightarrow 8$ ——

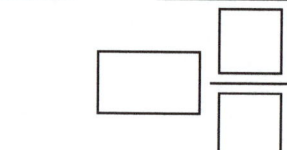

$- 2\frac{2}{6} \longrightarrow 2$ ——

B. $12\frac{1}{4} \longrightarrow 11$ ——

$- 7\frac{6}{8} \longrightarrow 7$ ——

C. $24\frac{1}{3} \longrightarrow 23$ ——

$-15\frac{2}{5} \longrightarrow 15$ ——

D. $31\frac{4}{5} \longrightarrow 31$ ——

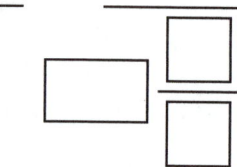

$-28\frac{3}{4} \longrightarrow 28$ ——

E. $6\frac{3}{5} \longrightarrow 5$ ——

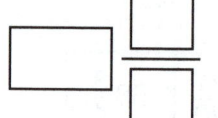

$- 2\frac{19}{20} \longrightarrow 2$ ——

F.

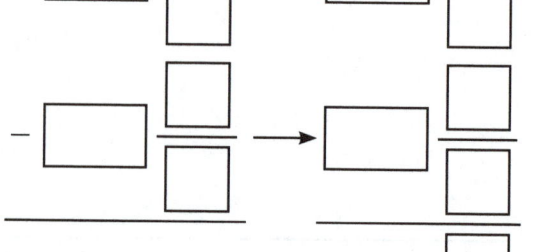

Practice

▶ **Find each difference. Write each answer in simplest form.**

1. $9\frac{1}{3}$
    $-\ 5\frac{1}{2}$

2. $14\frac{7}{8}$
    $-\ 6\frac{1}{4}$

3. $21\frac{1}{4}$
    $-\ 13\frac{11}{16}$

4. $40\frac{3}{5}$
    $-\ 28\frac{7}{8}$

5. $52\frac{5}{6}$
    $-\ 37\frac{1}{12}$

6. $71\frac{1}{2}$
    $-\ 49\frac{5}{8}$

7. $35\frac{2}{6}$
    $-\ 29\frac{2}{3}$

8. $31\frac{1}{4}$
    $-\ 21\frac{9}{12}$

9. $15\frac{3}{7}$
    $-\ 9\frac{4}{5}$

Complete this page with your teacher.

▶ **Find each product. Write each answer in simplest form.**

A. $\dfrac{5}{8} \times \dfrac{1}{3} =$ _____

$\dfrac{5}{8}$

$\dfrac{1}{3}$

B. $\dfrac{1}{2} \times \dfrac{4}{6} =$ _____ = _____

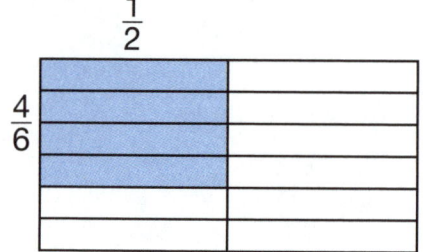

C. $\dfrac{2}{3} \times \dfrac{3}{4} =$ _____ = _____

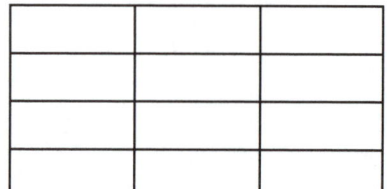

D. $\dfrac{1}{4} \times \dfrac{4}{5} =$ _____ = _____

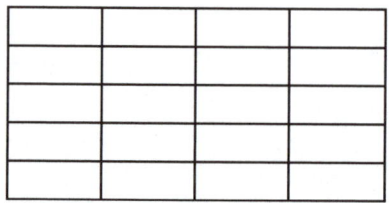

E. $\dfrac{3}{5} \times \dfrac{1}{3} =$ _____ = _____

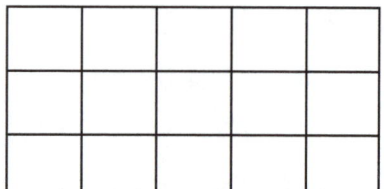

F. $\dfrac{5}{6} \times \dfrac{1}{2} =$ _____

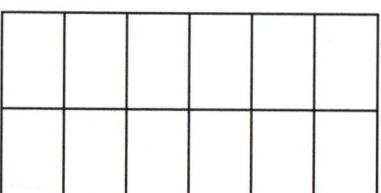

G. $\dfrac{\square}{\square} \times \dfrac{\square}{\square} = \dfrac{\square}{\square}$

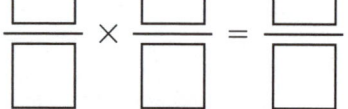

Practice

▶ **Find each product. Write each answer in simplest form.**

1. $\dfrac{1}{5} \times \dfrac{5}{6} =$ _____

2. $\dfrac{2}{7} \times \dfrac{3}{4} =$ _____

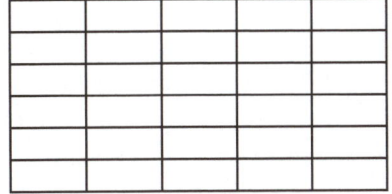   

3. $\dfrac{5}{8} \times \dfrac{2}{5} =$ _____

4. $\dfrac{2}{4} \times \dfrac{2}{3} =$ _____

5. $\dfrac{7}{9} \times \dfrac{6}{10} =$ _____

6. $\dfrac{3}{5} \times \dfrac{1}{2} =$ _____

7. $\dfrac{5}{6} \times \dfrac{2}{8} =$ _____

8. $\dfrac{7}{9} \times \dfrac{4}{5} =$ _____

9. $\dfrac{8}{10} \times \dfrac{5}{8} =$ _____

10. $\dfrac{9}{16} \times \dfrac{2}{3} =$ _____

11. $\dfrac{1}{4} \times \dfrac{3}{4} =$ _____

12. $\dfrac{8}{10} \times \dfrac{6}{7} =$ _____

13. $\dfrac{4}{5} \times \dfrac{9}{10} =$ _____

14. $\dfrac{3}{8} \times \dfrac{1}{6} =$ _____

Complete this page with your teacher.

▶ **Multiply to show that the numbers in each pair are reciprocals.**

A. $\frac{5}{6}$ and $\frac{6}{5}$ $\frac{5}{6} \times \frac{6}{5} = \frac{30}{30}$, or 1

B. 3 and $\frac{1}{3}$ $\frac{3}{1} \times$ _____ = _____, or 1

C. $2\frac{3}{4}$ and $\frac{4}{11}$ _____ × _____ = _____, or 1

▶ **Find the reciprocal of each number. Check your answer.**

D. $\frac{7}{8}$ and _____   _____ × _____ = _____, or 1

E. 9 and _____   _____ × _____ = _____, or 1

F. $3\frac{1}{5}$ and _____   _____ × _____ = _____, or 1

G. $\frac{2}{5}$ and _____   _____ × _____ = _____, or 1

H. 14 and _____   _____ × _____ = _____, or 1

I. $4\frac{4}{5}$ and _____   _____ × _____ = _____, or 1

J. _____ and _____   _____ × _____ = _____, or 1

Practice

▶ **Find the reciprocal of each number. Circle the correct answer.**

1. $1\frac{3}{8}$

   A. $\frac{4}{8}$ C. $\frac{8}{11}$

   B. $\frac{11}{8}$ D. $\frac{4}{11}$

2. 16

   A. $\frac{1}{6}$ C. $\frac{16}{1}$

   B. $\frac{1}{16}$ D. $\frac{6}{1}$

3. $2\frac{7}{10}$

   A. $\frac{27}{10}$ C. $\frac{20}{7}$

   B. $\frac{10}{27}$ D. $\frac{7}{27}$

4. $\frac{7}{12}$

   A. $1\frac{7}{5}$ C. $\frac{7}{12}$

   B. $1\frac{5}{12}$ D. $\frac{12}{7}$

5. $\frac{5}{9}$

   A. $1\frac{1}{4}$ C. $\frac{10}{9}$

   B. $1\frac{4}{9}$ D. $1\frac{4}{5}$

6. $5\frac{3}{4}$

   A. $\frac{23}{4}$ C. $\frac{4}{23}$

   B. $\frac{5}{4}$ D. $\frac{8}{5}$

7. 35

   A. $\frac{1}{35}$ C. $1\frac{3}{5}$

   B. $\frac{5}{3}$ D. $\frac{35}{10}$

8. $\frac{6}{14}$

   A. $1\frac{2}{6}$ C. $3\frac{1}{14}$

   B. $2\frac{1}{3}$ D. $1\frac{4}{6}$

9. $1\frac{5}{6}$

   A. $6\frac{1}{5}$ C. 31

   B. $\frac{6}{12}$ D. $\frac{6}{11}$

10. $5\frac{3}{9}$

    A. $\frac{9}{48}$ C. $\frac{9}{45}$

    B. $5\frac{9}{3}$ D. $\frac{45}{9}$

Complete this page with your teacher.

▶ **Find each quotient. Write each answer in simplest form.**

A. $\frac{1}{2} \div \frac{1}{4}$

$\frac{1}{2} \times \frac{4}{1} =$ _____ = _____

$\frac{1}{2}$	
$\frac{1}{2}$	$\frac{1}{2}$

$\frac{1}{4}$	$\frac{1}{4}$	$\frac{1}{4}$	$\frac{1}{4}$

B. $\frac{2}{3} \div \frac{1}{9}$

$\frac{2}{3} \times \frac{9}{1} =$ _____ = _____

$\frac{2}{3}$		
$\frac{1}{3}$	$\frac{1}{3}$	$\frac{1}{3}$

$\frac{1}{9}$	$\frac{1}{9}$	$\frac{1}{9}$	$\frac{1}{9}$	$\frac{1}{9}$	$\frac{1}{9}$	$\frac{1}{9}$	$\frac{1}{9}$	$\frac{1}{9}$

C. $\frac{5}{6} \div \frac{8}{10}$

_____ × _____ = _____

= _____

D. $\frac{2}{5} \div \frac{4}{7}$

_____ × _____ = _____

= _____

E. $\frac{3}{8} \div \frac{1}{5}$

_____ × _____ = _____

= _____

F. $\frac{9}{10} \div \frac{1}{2}$

_____ × _____ = _____

= _____

G. _____ ÷ _____

_____ × _____ = _____ = _____

Practice

▶ **Find each quotient. Write each answer in simplest form.**

1. $\dfrac{3}{16} \div \dfrac{1}{4} =$ _____

2. $\dfrac{2}{8} \div \dfrac{4}{5} =$ _____

3. $\dfrac{4}{5} \div \dfrac{2}{3} =$ _____

4. $\dfrac{3}{6} \div \dfrac{2}{9} =$ _____

5. $\dfrac{7}{10} \div \dfrac{3}{4} =$ _____

6. $\dfrac{5}{12} \div \dfrac{5}{6} =$ _____

7. $\dfrac{2}{5} \div \dfrac{2}{6} =$ _____

8. $\dfrac{2}{9} \div \dfrac{2}{9} =$ _____

9. $\dfrac{8}{12} \div \dfrac{3}{4} =$ _____

10. $\dfrac{5}{8} \div \dfrac{1}{2} =$ _____

11. $\dfrac{2}{7} \div \dfrac{1}{3} =$ _____

12. $\dfrac{9}{10} \div \dfrac{3}{5} =$ _____

13. $\dfrac{1}{3} \div \dfrac{1}{4} =$ _____

14. $\dfrac{5}{7} \div \dfrac{10}{15} =$ _____

15. $\dfrac{1}{8} \div \dfrac{1}{2} =$ _____

16. $\dfrac{2}{9} \div \dfrac{3}{12} =$ _____

17. $\dfrac{2}{3} \div \dfrac{5}{7} =$ _____

18. $\dfrac{13}{8} \div \dfrac{9}{27} =$ _____

Complete this page with your teacher.

▶ **Find each product.**

A. $2\frac{3}{5} \times 1\frac{1}{4} = \frac{13}{5} \times \frac{5}{4}$

      $= \frac{13 \times \cancel{5}^{1}}{\cancel{5}_{1} \times 4}$

      $= \frac{13}{4} = $ _____

- Write mixed numbers as improper fractions.

- Cancel common factors.

- Write the product in simplest form.

B. $4\frac{2}{3} \times 2\frac{5}{8} = $ _____

C. $1\frac{8}{10} \times 1\frac{2}{9} = $ _____

D. $5\frac{1}{3} \times 2\frac{3}{4} = $ _____

E. $6\frac{4}{5} \times 3\frac{1}{2} = $ _____

▶ **Find each quotient.**

F. $3\frac{1}{3} \div 1\frac{2}{5} = \frac{10}{3} \div \frac{7}{5}$

      $= \frac{10}{3} \times \frac{5}{7}$

      $= \frac{50}{21} = $ _____

- Write mixed numbers as improper fractions.

- Multiply by the reciprocal.

- Write the quotient in simplest form.

G. $5\frac{1}{2} \div 2\frac{1}{8} = $ _____

H. $2\frac{4}{6} \div 1\frac{1}{3} = $ _____

I. $7\frac{2}{9} \div 5\frac{2}{3} = $ _____

J. $1\frac{1}{8} \div 4\frac{5}{9} = $ _____

K.   _____ ◯ _____ = _____

Practice

▶ **Find each product or quotient. Show your work. Then shade the matching answer in the table below.**

1. $4\frac{7}{8} \div 2\frac{1}{4} =$ _____

2. $3\frac{5}{9} \times 2\frac{5}{8} =$ _____

3. $3\frac{1}{3} \times 2\frac{2}{5} =$ _____

4. $5\frac{1}{2} \div 2\frac{9}{10} =$ _____

5. $6\frac{1}{4} \div 2\frac{7}{8} =$ _____

6. $4\frac{1}{5} \times 2\frac{1}{7} =$ _____

7. $1\frac{8}{10} \times 2\frac{4}{6} =$ _____

8. $8\frac{1}{10} \div 1\frac{4}{5} =$ _____

$2\frac{4}{23}$	9	$4\frac{4}{5}$	$2\frac{1}{6}$	$1\frac{26}{29}$	8	$9\frac{1}{3}$	$4\frac{1}{2}$

Complete this page with your teacher.

▶ Write each fraction as a terminating or repeating decimal.
Use a calculator to check your answer.

A. $\frac{3}{4}$ = _____

```
      0.  7  □
   4) 3.  0  0
    − 2  8  ↓
      ┌──┬──┐
      │□ │□ │
      └──┴──┘
    − □  □
      ─────
      □
```

B. $\frac{1}{6}$ = _____

```
      0.  1  □  □  ...
   6) 1.  0  0  0
    − □  ↓
      ┌──┬──┐
      │□ │□ │
      └──┼──┤
    − □  □  ↓
      ┌──┬──┐
      │□ │□ │
      └──┴──┘
```

C. $\frac{12}{25}$ = _____   ⟌‾‾‾

D. _____ = _____   ⟌‾‾‾

▶ Use a calculator to find the equivalent decimal for each fraction.
Write T for each terminating decimal and R for each repeating decimal.

E. $\frac{5}{8}$ _____ , _____

F. $\frac{2}{9}$ _____ , _____

G. $\frac{15}{18}$ _____ , _____

H. _____ _____ , _____
     fraction decimal

Practice

▶ **Write each fraction as a terminating or repeating decimal. Use a calculator to check your answer.**

1. $\frac{5}{12} =$ _____

2. $\frac{3}{8} =$ _____

3. $\frac{12}{20} =$ _____

4. $\frac{8}{15} =$ _____

▶ **Use a calculator to find the equivalent decimal for each fraction. Write T for each terminating decimal and R for each repeating decimal.**

5. $\frac{2}{3}$ $0.\overline{6}$, R

6. $\frac{1}{2}$ _____ , _____

7. $\frac{5}{6}$ _____ , _____

8. $\frac{1}{9}$ _____ , _____

9. $\frac{4}{5}$ _____ , _____

10. $\frac{3}{20}$ _____ , _____

11. $\frac{14}{18}$ _____ , _____

12. $\frac{7}{12}$ _____ , _____

13. $\frac{6}{24}$ _____ , _____

Complete this page with your teacher.

▶ **Find each product. Show your work.**

A.

```
        2 . 7   8      ← 2 decimal places
    ×     8 . 5        ← 1 decimal place
```

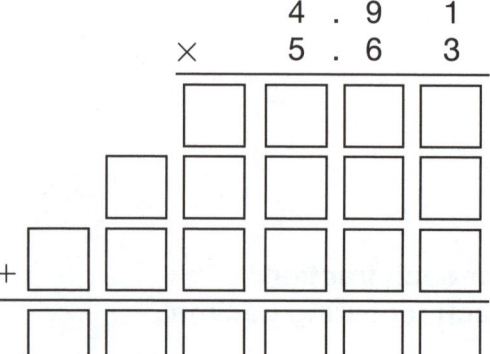

← 3 decimal places

B.

```
        4 . 9   1      ← 2 decimal places
    ×   5 . 6   3      ← 2 decimal places
```

← 4 decimal places

C.

```
        7 . 0   8
    ×     9 . 4
```

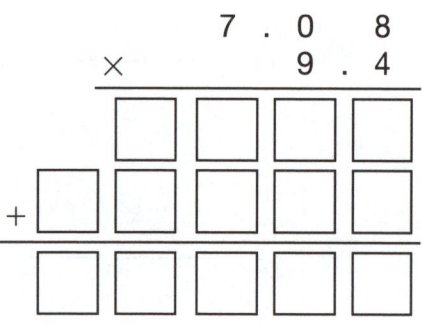

D.

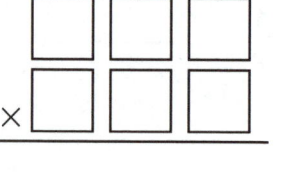

Practice

▶ **Find each product.**

1. 2.98
    × 4.6

2. 70.2
    × 6.8

3. 3.51
    × 9.24

4. 8.37
    × 2.7

5. 50.45
    × 3.89

6. 67.3
    × 5.4

7. 160.1
    × 7.05

8. 9.2
    × 1.9

9. 4.48
    × 5.6

10. 13.7
    × 6.8

11. 8.45
    × 27.6

12. 7.09
    × 8.3

13. 8.92
    × 6.4

14. 3.76
    × 5.4

15. 84.5
    × 6.8

Complete this page with your teacher.

▶ **Find each quotient. Show your work.**

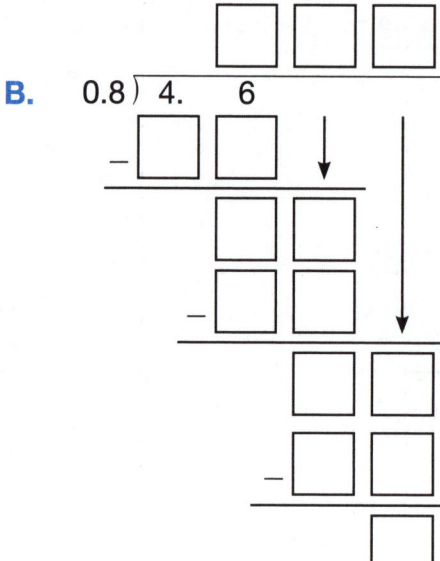

A. $0.4\overline{)3.74}$ ⟶ $4\overline{)37.40}$

```
        9. □ □
  4)3 7. 4  0
  − 3 6
     □ □
     □ □
   − □ □
       □ □
       □ □
     − □ □
         □
```

B. $0.8\overline{)4.6}$

C. $0.4\overline{)3.06}$

D. $2.6\overline{)13.26}$

E. $0.3\overline{)64.8}$

F.

Practice

▶ **Find each quotient.**

1. $0.5\overline{)23.6}$

2. $0.7\overline{)1.89}$

3. $0.9\overline{)5.346}$

4. $0.4\overline{)14.6}$

5. $0.2\overline{)3.71}$

6. $0.3\overline{)0.48}$

7. $0.8\overline{)20.7}$

8. $0.6\overline{)3.75}$

9. $0.5\overline{)2.15}$

10. $1.4\overline{)13.3}$

11. $3.9\overline{)20.475}$

12. $6.13\overline{)49.04}$

Complete this page with your teacher.

▶ **Write each ratio in simplest form.**

A.

stars to squares: 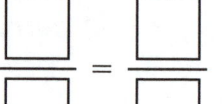 ← first term

← second term

squares to stars: 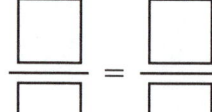 ← first term

← second term

B.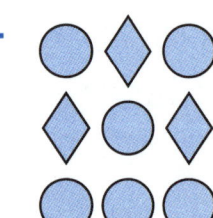

circles to diamonds: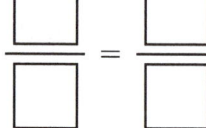

diamonds to circles: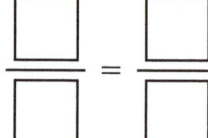

C.

_____ to _____: ☐/☐

D. A bag contains 8 cherry, 10 lemon, 9 orange, and 12 lime jellybeans. What is the ratio of lemon to lime jellybeans?

☐/☐ = ☐/☐

E. A bag holds _____ red, _____ blue, and _____ green marbles.

What is the ratio of _____ to _____ marbles?

☐/☐

Practice

▶ **Write each ratio in simplest form.**

1.

squares to triangles:

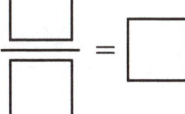
$$\frac{}{} = \square$$

triangles to squares:

$$\frac{}{} = \frac{}{}$$

2.

stars to circle:

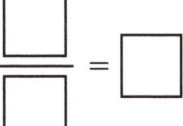
$$\frac{}{} = \square$$

circle to stars:

$$\frac{}{}$$

3. A piggy bank holds 18 quarters, 10 dimes, 6 nickels, and 15 pennies. What is the ratio of pennies to quarters?

$$\frac{}{} = \frac{}{}$$

4. A bag contains 7 oatmeal raisin, 12 chocolate chip, and 9 plain granola bars. What is the ratio of chocolate chip granola bars to the total number of granola bars?

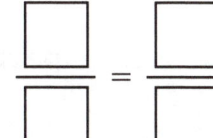
$$\frac{}{} = \frac{}{}$$

5. Emiko's math class has 25 students. There are 10 male students. What is the ratio of male students to female students?

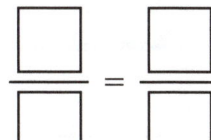
$$\frac{}{} = \frac{}{}$$

Complete this page with your teacher.

▶ **Solve each problem.**

A. Students at a fundraiser made $625 in 5 hours. How much money did they raise per hour?

$_____ per hour

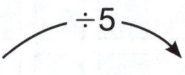

_____ = _____
                 1

B. Corinne ran 84 miles in 7 weeks. How many miles did she run each week?

_____ miles per week

_____ = _____
                 1

C. Four pounds of oranges cost $4.60. How much does 1 pound of oranges cost?

$_____ per pound

_____ = _____

D. A family traveled 520 miles in 8 hours. What was their average speed in miles per hour?

_____ = _____

E. _____

_____

_____   _____ = _____

_____

Practice

▶ **Solve each problem. Show your work.**

1. A student types 124 words in 4 minutes.
   How many words does the student type per minute?

   _____ words per minute

2. Six T-shirts cost $55.20.
   What is the unit rate for the T-shirts?

   $_____

3. A family drove 828 miles in 4 days.
   How many miles did they drive per day?

   _____ miles per day

4. One night, 7.5 inches of snow fell in 3 hours.
   How much snow fell per hour?

   _____ inches per hour

5. A store is offering 3 different sales on socks. Find each offer's
   unit rate and determine which sale offer is the best buy.

Sale	Quantity	Cost	Unit Rate
Orange Tag	2-pack	$3.58	
Green Tag	3-pack	$5.07	
Red Tag	5-pack	$8.75	

   The _____ sale is the best buy.

Complete this page with your teacher.

▶ **Use a ratio table to solve each problem.**

A. A store sells 12 cans of cat food for $5.88. How much will it cost to buy 2 cans of cat food?

$$÷12 \qquad ×2$$

Cost in Dollars	5.88		
Cans of Cat Food	12		2

$$÷12 \qquad ×2$$

It will cost $_____ to buy 2 cans of cat food.

B. A student can run 8 miles in 48 minutes. How long will it take the student to run 3 miles?

Total Minutes	48		
Miles	8		3

The student can run 3 miles in _____ minutes.

C. _____

_____

_____

Practice

▶ **Use a ratio table to solve each problem.**

1. A car travels 392 miles on 14 gallons of gas. How many miles can it travel on 8 gallons of gas?

   _____ miles

Number of Miles			
Gallons of Gas			

2. A student reads 180 pages in 4 days. How many pages can the student read in 3 days?

   _____ pages

Number of Pages			
Number of Days			

3. Ten pounds of potatoes sell for $6.70. How much will it cost to buy 7 pounds of potatoes?

   $_____

Cost in Dollars			
Pounds of Potatoes			

4. Antonio can deliver 105 newspapers in 3 hours. How many newspapers can he deliver in 2 hours?

   _____ newspapers

Number of Newspapers			
Number of Hours			

Complete this page with your teacher.

▶ **Simplify each ratio to determine whether the ratios form a proportion.**

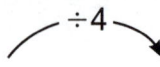

A. $\dfrac{96}{4} \overset{?}{=} \dfrac{120}{5}$

$\dfrac{96}{4} = $ _____

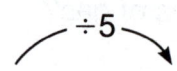

$\dfrac{120}{5} = $ _____

Do the ratios form a proportion? _____

▶ **Use cross products to determine whether the ratios form a proportion.**

B. $\dfrac{132}{12} \overset{?}{=} \dfrac{108}{9}$

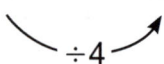

$132 \times 9 \overset{?}{=} 12 \times 108$

_____ $\overset{?}{=}$ _____

Do the ratios form a proportion? _____

▶ **Simplify or use cross products to determine whether the ratios form a proportion. Circle your answer.**

C. $\dfrac{114}{3} \overset{?}{=} \dfrac{490}{14}$ Yes   No

D. $\dfrac{80}{16} \overset{?}{=} \dfrac{65}{13}$ Yes   No

E. $\dfrac{29}{45} \overset{?}{=} \dfrac{145}{180}$ Yes   No

F. $\dfrac{50}{23} \overset{?}{=} \dfrac{400}{184}$ Yes   No

G. _____ $\overset{?}{=}$ _____   Yes   No

Practice

▶ **Simplify or use cross products to determine whether the ratios form a proportion. Circle your answer.**

1. $\frac{180}{15} \overset{?}{=} \frac{216}{18}$ Yes
No

2. $\frac{480}{20} \overset{?}{=} \frac{240}{12}$ Yes
No

3. $\frac{153}{9} \overset{?}{=} \frac{195}{13}$ Yes
No

4. $\frac{75}{225} \overset{?}{=} \frac{35}{105}$ Yes
No

5. $\frac{72}{36} \overset{?}{=} \frac{94}{52}$ Yes
No

6. $\frac{84}{8} \overset{?}{=} \frac{147}{14}$ Yes
No

7. $\frac{25}{315} \overset{?}{=} \frac{30}{378}$ Yes
No

8. $\frac{51}{17} \overset{?}{=} \frac{116}{29}$ Yes
No

9. $\frac{448}{16} \overset{?}{=} \frac{756}{27}$ Yes
No

10. $\frac{4}{124} \overset{?}{=} \frac{7}{217}$ Yes
No

11. $\frac{4}{21} \overset{?}{=} \frac{300}{1,575}$ Yes
No

12. $\frac{17}{34} \overset{?}{=} \frac{1}{3}$ Yes
No

Complete this page with your teacher.

▶ **Write a percent to represent each model.**

A.

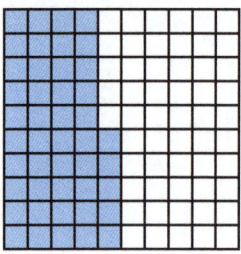

_____ out of 100 = _____%

B.

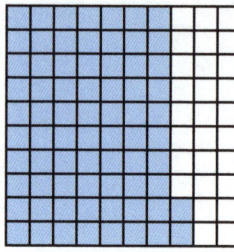

_____ out of 100 = _____%

C.

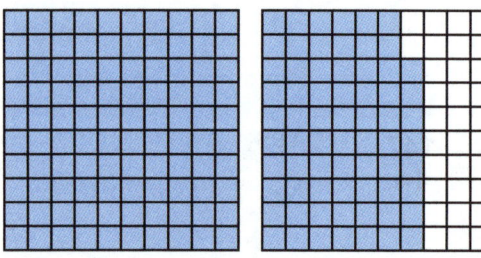

1 whole + _____ out of 100 = _____%

D.

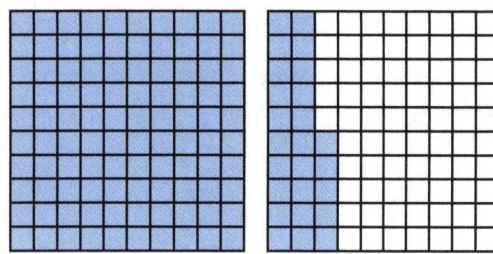

1 whole + _____ out of 100 = _____%

E.

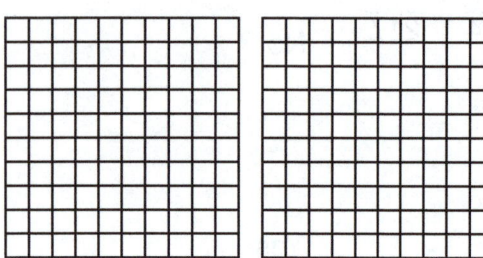

= _____%

Practice

▶ **Write a percent to represent each model.**

1.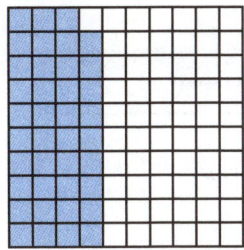

_____%

2.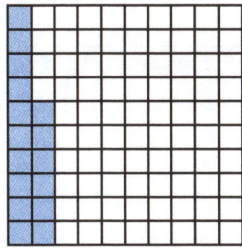

_____%

3.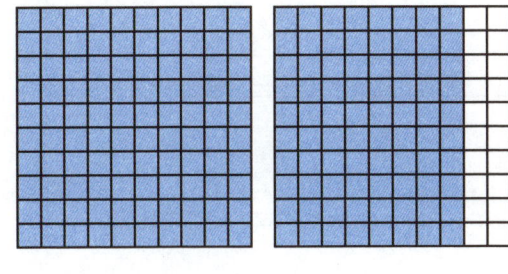

_____%

4.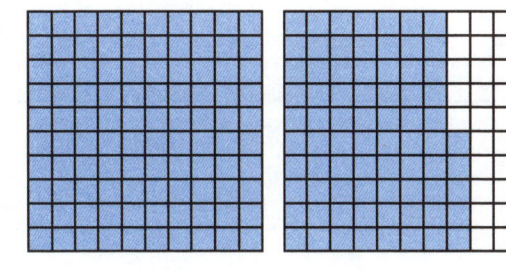

_____%

5.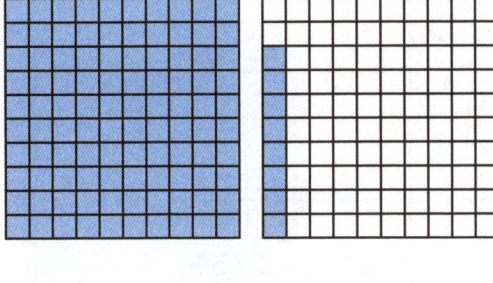

_____%

6.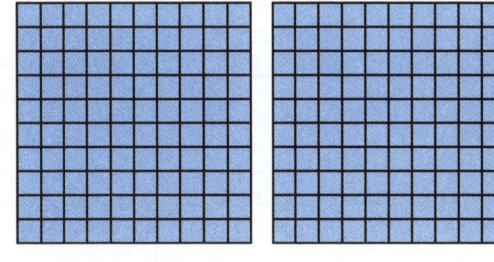

_____%

7.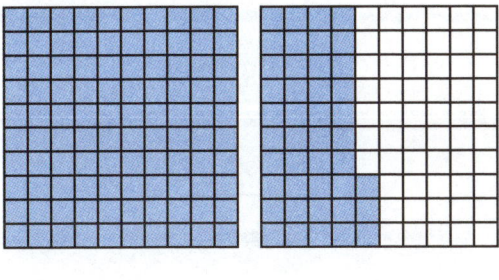

_____%

8.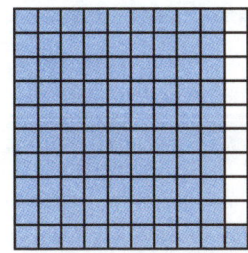

_____%

Complete this page with your teacher.

▶ Write the equivalent forms of each value.

	Fraction	Decimal	Percent
A.	$\frac{1}{5} = \frac{20}{100}$	$0.2 = 0.20$	20%
B.	_____	_____	50%
C.	$\frac{7}{20}$	_____	_____
D.	_____	0.875	_____
E.	_____	_____	25%
F.	$\frac{11}{50}$	_____	_____
G.	_____	0.625	_____
H.	_____	_____	37.5%

	Fraction	Decimal	Percent
I.	_____	_____	_____

Practice

▶ **Write the equivalent forms of each value.**

	Fraction	Decimal	Percent
1.	_____	_____	80%
2.	$\frac{1}{8}$	_____	_____
3.	_____	0.04	_____
4.	$\frac{9}{24}$	_____	_____
5.	_____	_____	55%
6.	_____	0.25	_____
7.	$\frac{37}{50}$	_____	_____
8.	_____	0.6	_____
9.	_____	_____	42%

Complete this page with your teacher.

▶ **Find 35% of 160.**

A. Write the percent as a decimal.

$35\% = 0.35$

$0.35 \times 160 =$ _____

B. Write the percent as a fraction.

$\div 5$

$35\% = \dfrac{35}{100} = \dfrac{\boxed{}}{\boxed{}}$

$\div 5$

$\dfrac{\boxed{}}{\boxed{}} \times 160 = \dfrac{\boxed{}}{\boxed{}} =$ _____

▶ **Find 80% of 40.**

C. Write the percent as a decimal.

$80\% =$ _____

_____ $\times 40 =$ _____

D. Write the percent as a fraction.

$80\% = \dfrac{\boxed{}}{\boxed{}} = \dfrac{\boxed{}}{\boxed{}}$

$\dfrac{\boxed{}}{\boxed{}} \times 40 = \dfrac{\boxed{}}{\boxed{}} =$ _____

Find _____% of _____.

E. Write the percent as a decimal.

_____% = _____

_____ × _____ = _____

F. Write the percent as a fraction.

_____% = _____ = _____

_____ × _____ = _____

Practice

▶ **Write each percent as a decimal to find the percent of each number.**

1. 90% of 240   90% = _____

_____ × _____ = _____

2. 32% of 450   32% = _____

_____ × _____ = _____

3. 45% of 80   45% = _____

_____ × _____ = _____

4. 2% of 500   2% = _____

_____ × _____ = _____

5. 18% of 150   18% = _____

_____ × _____ = _____

6. 250% of 16   250% = _____

_____ × _____ = _____

▶ **Write each percent as a fraction to find the percent of each number.**

7. 70% of 70

70% = _____ = _____

_____ × _____ = _____

8. 5% of 20

5% = _____ = _____

_____ × _____ = _____

9. 165% of 140

165% = _____ = _____

_____ × _____ = _____

10. 59% of 300

59% = _____

_____ × _____ = _____

11. 325% of 60

325% = _____ = _____

_____ × _____ = _____

12. 200% of 48

200% = _____ = _____

_____ × _____ = _____

Complete this page with your teacher.

▶ **Write >, <, or = to compare the values.**

A. $\frac{14}{5}$ ◯ 195%

$\frac{14}{5} = 2\frac{4}{5} = 2.80 = 280\%$, so $\frac{14}{5}$ ⟩ 195%

B. 1.4 ◯ $\frac{7}{4}$

$\frac{7}{4} = 1\frac{\square}{\square} = 1\frac{\square}{\square}$ = _____, so 1.4 ◯ $\frac{7}{4}$

C. _____ ◯ _____

▶ **Order the values from least to greatest.**

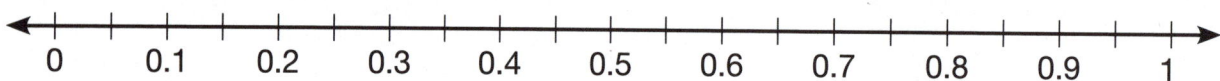

D. $\frac{1}{2}$, 0.2, 35%   _____ < _____ < _____

E. 0.7, 85%, $\frac{11}{20}$ _____ < _____ < _____

F. _____ , _____ , _____   _____ < _____ < _____

Practice

Goal: Compare and order values in various forms.

▶ **Write >, <, or = to compare the values.**

1. $\frac{3}{8}$ ◯ 35%

2. $\frac{3}{2}$ ◯ 1.25

3. 2.8 ◯ 300%

4. $\frac{1}{25}$ ◯ 4%

5. 115% ◯ 1.2

6. 430% ◯ $\frac{23}{5}$

7. $\frac{9}{10}$ ◯ 8%

8. $\frac{3}{20}$ ◯ 15%

▶ **Order the values from least to greatest.**

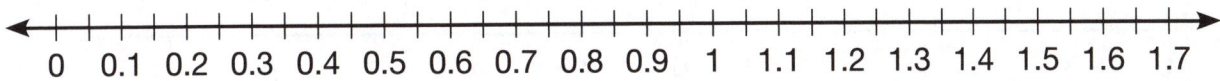

```
0   0.1  0.2  0.3  0.4  0.5  0.6  0.7  0.8  0.9  1   1.1  1.2  1.3  1.4  1.5  1.6  1.7
```

9. 165%, $\frac{5}{4}$, 1.5   _____ < _____ < _____

10. 60%, $\frac{4}{5}$, 0.7   _____ < _____ < _____

11. $\frac{9}{10}$, 140%, 1.25   _____ < _____ < _____

12. 0.5, 8%, $\frac{12}{8}$ _____ < _____ < _____

13. 0.62, $\frac{5}{6}$, 75%   _____ < _____ < _____

14. $\frac{3}{5}$, 0.59, 61%   _____ < _____ < _____

15. 0.99, 100%, $\frac{9}{10}$ _____ < _____ < _____

16. $\frac{1}{2}$, $\frac{3}{8}$, 45%   _____ < _____ < _____

Complete this page with your teacher.

▶ **Use the number line to compare the integers. Write >, <, or =.**

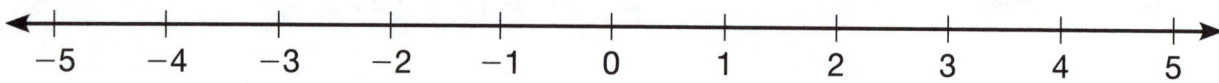

A. −5 ◯ 3   B. −2 ◯ −3

C. 4 ◯ −1   D. −4 ◯ 0

E. −1 ◯ −5   F. 0 ◯ −3

▶ **Order the integers from least to greatest.**

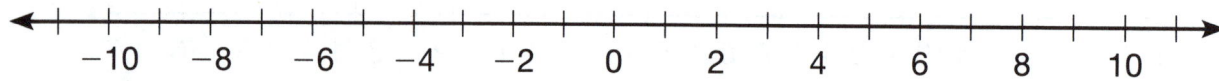

G. −3, −8, 7, 0, −1   H. 10, −2, 4, −6, −5

_____   _____

I. 4, −6, 5, 0, −8   J. 7, 9, −1, 1, 0

_____   _____

K. _____ ◯ _____   L. _____

                            _____

Practice

▶ **Use the number line to compare the integers. Write >, <, or =.**

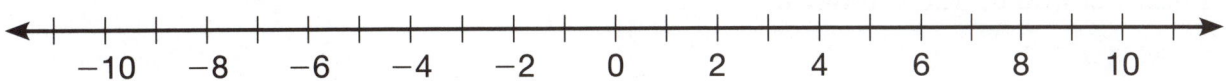

1. −9 ◯ −7   2.   3 ◯ −1

3.   0 ◯ −6   4. −5 ◯ −2

5. −8 ◯ −4   6. −3 ◯ −3

7. 10 ◯ −10   8. −7 ◯ −8

9. −2 ◯ −3   10. −10 ◯ −7

▶ **Use the number line above to order the integers from least to greatest.**

11. 9, −6, 2, −3, 0   _____

12. −7, 5, −1, 4, −6   _____

13. −5, −8, 1, −10, −2   _____

14. 6, −4, 3, 7, −9   _____

15. −3, −6, 3, −8, 0   _____

16. 1, −9, −2, 2, 4   _____

Complete this page with your teacher.

▶ **Use a number line to add integers.**

A. −2 + 5 = _____

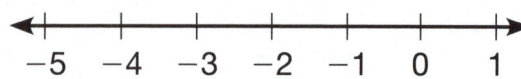

B. −1 + (−3) = _____

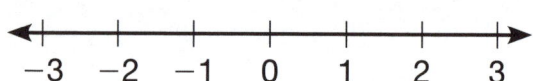

C. 1 + (−4) = _____

D. −3 + 4 = _____

▶ **Use the number line to find each sum.**

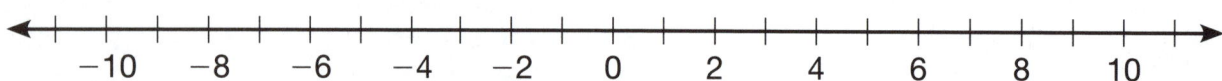

E. −4 + (−7) = _____

F. 2 + 8 = _____

G. 5 + (−9) = _____

H. −3 + (−3) = _____

I. −10 + 7 = _____

J. −8 + 13 = _____

K. 6 + (−9) = _____

L. −1 + 2 = _____

M. _____ + _____ = _____

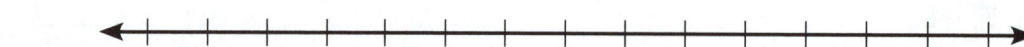

Practice

▶ **Use the number line to find each sum.**

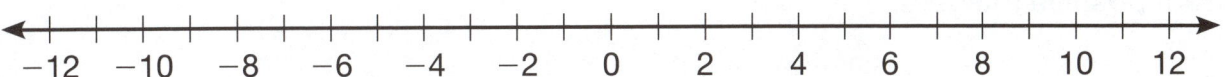

$$-12 \quad -10 \quad -8 \quad -6 \quad -4 \quad -2 \quad 0 \quad 2 \quad 4 \quad 6 \quad 8 \quad 10 \quad 12$$

1. $-4 + (-7) =$ _____

2. $2 + 8 =$ _____

3. $5 + (-9) =$ _____

4. $-3 + (-3) =$ _____

5. $-10 + 7 =$ _____

6. $-8 + 13 =$ _____

7. $0 + (-5) =$ _____

8. $-6 + (-4) =$ _____

9. $4 + (-2) =$ _____

10. $-12 + 7 =$ _____

11. $6 + (-1) =$ _____

12. $10 + (-1) =$ _____

13. $-5 + (-7) =$ _____

14. $2 + (-9) =$ _____

▶ **Write an addition problem for each situation. Then find the sum.**

15. The temperature outside is 6°F. If the temperature drops 10° overnight, what will the low temperature be for the night?

    _____ + _____ = _____°F

16. A diver swims 7 feet below the water's surface. Then the diver swims another 8 feet below the surface. What is the location of the diver?

    _____ + _____ = _____ feet

17. In a football game, a team advances 4 yards on their first play. On their second play, the team loses 9 yards. What integer represents the team's change in position after the two plays?

    _____ + _____ = _____ yards

Complete this page with your teacher.

▶ **Subtract positive integers.**

A. $6 - 2 =$ _____

$6 + (-2) =$ _____

0 1 2 3 4 5 6

B. $-4 - 1 =$ _____

$-4 +$ _____ $=$ _____

C. $3 - 9 =$ _____

$3 +$ _____ $=$ _____

▶ **Subtract negative integers.**

D. $-7 - (-3) =$ _____

$-7 + (+3) =$ _____

−8 −7 −6 −5 −4 −3 −2 −1 0

E. $-1 - (-6) =$ _____

$-1 +$ _____ $=$ _____

F. $5 - (-3) =$ _____

$5 +$ _____ $=$ _____

G. _____ − _____ $=$ _____

Practice

▶ **Find each difference. Then write the letter of the matching difference on the line.**

_____ **1.** $-2 - 8 =$ **A.** -4

_____ **2.** $-9 - (-6) =$ **B.** 9

_____ **3.** $3 - 7 =$ **C.** -10

_____ **4.** $-4 - (-11) =$ **D.** 3

_____ **5.** $-5 - (-8) =$ **E.** -7

_____ **6.** $-6 - (-10) =$ **F.** -3

_____ **7.** $7 - (-2) =$ **G.** -1

_____ **8.** $-4 - 3 =$ **H.** 7

_____ **9.** $8 - 9 =$ **I.** 5

_____ **10.** $-3 - (-8) =$ **J.** 4

Complete this page with your teacher.

If the factors have different signs, the product is negative.

$-5 \times 3 = -15$
$7 \times -5 = -35$

If the factors have the same sign, the product is positive.

$4 \times 9 = 36$
$-8 \times -2 = 16$

▶ **Find each product.**

A. $4 \times (-5) =$ _____   The factors have different signs, so the product is negative.

B. $-3 \times (-7) =$ _____   **C.** $-5 \times (-3) =$ _____

D. $6 \times 9 =$ _____   **E.** $-10 \times 8 =$ _____

F. $-5 \times (-9) =$ _____   **G.** $-9 \times (-7) =$ _____

H. $16 \times 19 =$ _____   **I.** $-5 \times 3 =$ _____

J. $21 \times -4 =$ _____   **K.** $-7 \times (-7) =$ _____

L. $14 \times (-3) =$ _____   **M.** $-20 \times (-3) =$ _____

N. _____ $\times$ _____ = _____

Practice

▶ **Find each product.**

1. $-7 \times 9 =$ _____

2. $8 \times (-3) =$ _____

3. $-5 \times (-6) =$ _____

4. $4 \times (-7) =$ _____

5. $8 \times (-11) =$ _____

6. $12 \times 6 =$ _____

7. $-14 \times 9 =$ _____

8. $-8 \times 8 =$ _____

9. $15 \times 5 =$ _____

10. $9 \times (-3) =$ _____

11. $-6 \times 8 =$ _____

12. $-16 \times (-7) =$ _____

13. $-10 \times 10 =$ _____

14. $7 \times (-4) =$ _____

15. $-5 \times (-14) =$ _____

16. $-12 \times 3 =$ _____

▶ **Write a multiplication sentence for each situation. Then find the product.**

17. A trivia game takes away 3 points for every incorrect answer. A player answers the first 5 questions incorrectly. What is this player's score so far?

_____

18. A diver descends 12 feet below sea level per minute. After 7 minutes, where is the diver compared to the water's surface?

_____

Complete this page with your teacher.

If the signs are different, the quotient is negative.

$-54 \div 9 = -6$

$-6 \times 9 = -54$

$16 \div (-4) = -4$

$-4 \times -4 = 16$

If the signs are the same, the quotient is positive.

$-24 \div (-8) = 3$

$3 \times (-8) = -24$

$72 \div 9 = 8$

$8 \times 9 = 72$

▶ **Find each quotient. Write a multiplication sentence to prove each answer is correct.**

A. $36 \div (-4) =$ _____

_____ × _____ = _____

B. $-42 \div (-6) =$ _____

_____ × _____ = _____

C. $-80 \div 5 =$ _____

_____ × _____ = _____

D. $75 \div 3 =$ _____

_____ × _____ = _____

E. $-56 \div (-7) =$ _____

_____ × _____ = _____

F. $108 \div (-9) =$ _____

_____ × _____ = _____

G. $49 \div (-7) =$ _____

_____ × _____ = _____

H. $-625 \div (-25) =$ _____

_____ × _____ = _____

I. $585 \div (-13) =$ _____

_____ × _____ = _____

J. _____ ÷ _____ = _____

_____ × _____ = _____

Practice

▶ **Find each quotient. Write a multiplication sentence to prove each answer is correct.**

1. 51 ÷ (−3) = _____

_____

2. −63 ÷ 7 = _____

_____

3. −32 ÷ (−4) = _____

_____

4. 26 ÷ (−2) = _____

_____

5. −49 ÷ 7 = _____

_____

6. 64 ÷ 4 = _____

_____

7. 60 ÷ (−5) = _____

_____

8. −45 ÷ (−9) = _____

_____

9. −84 ÷ 4 = _____

_____

10. 56 ÷ (−7) = _____

_____

11. 72 ÷ 2 = _____

_____

12. −12 ÷ (−12) = _____

_____

13. 40 ÷ (−10) = _____

_____

14. 30 ÷ (−5) = _____

_____

Complete this page with your teacher.

▶ **Absolute value is the distance of a number from 0 on a number line. It is always a positive value.**

$|-4| = 4$

$|4| = 4$

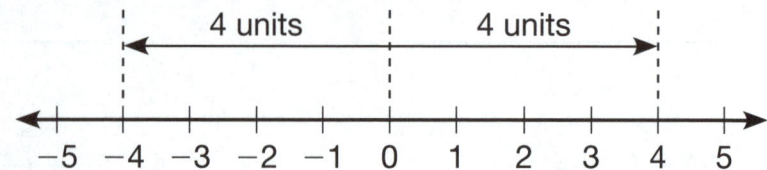

4 units   4 units

−5 −4 −3 −2 −1 0 1 2 3 4 5

▶ **Find the absolute value of each integer.**

A. $|-9| =$ _____

B. $|7| =$ _____

C. $|-6| =$ _____

▶ **Simplify each expression.**

D. $|-8| + |2| =$ _____

E. $|-10| - |-3| =$ _____

F. $|-5| \times |9| =$ _____

G. $|36| \div |-6| =$ _____

H. $|-6| + |5| =$ _____

I. $|-20| - |14| =$ _____

J. $|-2| \times |23| =$ _____

K. $|60| \div |-1| =$ _____

L. $\left|\ \boxed{}\ \right| =$ _____

M. $\left|\ \boxed{}\ \right| \bigcirc \left|\ \boxed{}\ \right| =$ _____

Practice

▶ **Find the absolute value of each integer.**

1. $|-12| =$ _____

2. $|18| =$ _____

3. $|-47| =$ _____

4. $|35| =$ _____

5. $|-20| =$ _____

6. $|-9| =$ _____

▶ **Simplify each expression.**

7. $|-7| + |-5| =$ _____

8. $|-19| - |4| =$ _____

9. $|-8| - |3| =$ _____

10. $|-42| \div |-7| =$ _____

11. $|12| \times |-9| =$ _____

12. $|-36| + |-14| =$ _____

13. $|-15| + |26| =$ _____

14. $|18| \times |-4| =$ _____

15. $|-17| - |8| =$ _____

16. $|13| + |-11| =$ _____

17. $|-51| + |-62| =$ _____

18. $|81| \times |-3| =$ _____

19. $|-71| - |30| =$ _____

20. $|31| + |-21| =$ _____

Complete this page with your teacher.

The **base** tells the number to be multiplied.

base $\longrightarrow 6^2 \longleftarrow$ exponent

The **exponent** tells how many times to multiply the base.

▶ **Simplify each exponent expression.**

A. $4^2 = $ _____ × _____ = _____

B. $5^3 = $ _____ × _____ × _____ = _____

C. $2^4 = $ _____ × _____ × _____ × _____ = _____

D. $\square^{\square} = $ _____ = _____

▶ **Simplify the exponent expressions. Then order them from least to greatest.**

E. $5^2, 4^3, 2^5, 3^4$ _____

$5^2 = $ _____ $4^3 = $ _____

$2^5 = $ _____ $3^4 = $ _____

F. _____ , _____ , _____ , _____   _____

Practice

▶ **For each statement, write T for *true* or F for *false*.**

_____ **1.** 3^2 means 3×3, or 9.

_____ **2.** 2^3 has a greater value than 3^2.

_____ **3.** In 7^4, the base is 4 and the exponent is 7.

_____ **4.** 5^3 means $5 \times 5 \times 5$, or 125.

_____ **5.** 6^2 means $2 \times 2 \times 2 \times 2 \times 2 \times 2$, or 64.

_____ **6.** 4^3 is equal to 8^2.

_____ **7.** In 9^3, the base is 9 and the exponent is 3.

▶ **Simplify the exponent expressions. Then order them from least to greatest.**

8. $2^2, 3^3, 4^3, 5^2$ _____

9. $6^2, 2^5, 3^2, 4^2$ _____

10. $3^4, 4^3, 2^5, 3^3$ _____

11. $2^2, 5^2, 3^2, 4^2$ _____

12. $2^5, 3^4, 4^3, 5^2$ _____

Complete this page with your teacher.

▶ **Write the prime factorization of each number.**

A.

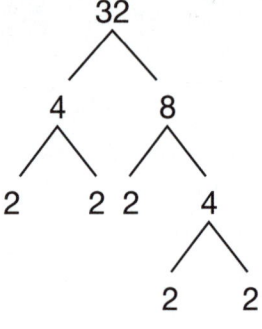

32

4   8

2   2 2   4

2   2

$32 = 2 \times 2 \times 2 \times 2 \times 2$, or 2^5

- Draw a factor tree for 32.

- Choose any pair of whole number factors to start.

- Continue factoring until all the factors are prime.

- Write 32 as the product of its prime factors.

B.

54

$54 = $ _____ ,

or _____

C.

90

$90 = $ _____ ,

or _____

D. _____ = _____ , _____

Practice

► **Write the prime factorization of each number. Then circle the correct matching answer.**

1. 72
  A. $2^2 \times 3^3$
  B. 8×3^3
  C. $2^2 \times 3 \times 4$
  D. $2^3 \times 3^2$

2. 81
  A. 2^8
  B. 3^4
  C. 3^3
  D. 9^3

3. 315
  A. $3 \times 5 \times 7$
  B. $3^2 \times 7$
  C. $3^2 \times 5 \times 7$
  D. $5^2 \times 7$

4. 100
  A. $2^2 \times 5^2$
  B. 4×2^5
  C. 2×5^2
  D. $2^2 \times 5$

5. 64
  A. 2^7
  B. 2^6
  C. 2^4
  D. 2^3

6. 125
  A. 2×5^2
  B. 3×5^3
  C. 3×5^2
  D. 5^3

7. 98
  A. 2×7^2
  B. 3×7^2
  C. $2^2 \times 7^2$
  D. $3^2 \times 7$

8. 216
  A. $2^2 \times 3^3$
  B. $2^3 \times 3^2$
  C. $2^3 \times 3^3$
  D. $2^4 \times 3^2$

Complete this page with your teacher.

▶ **Use the order of operations to simplify each expression.**

A. $12 + (8 \times 3) - 4^2$

$12 + (\boxed{}) - 4^2$

$12 + \boxed{} - \boxed{}$

$\boxed{} - \boxed{} = \underline{}$

> **Order of Operations**
>
> ( )   Parentheses
>
> 3^2 Exponents
>
> Then, from left to right:
>
> $\times, \div$ Multiply and divide.
>
> $+, -$ Add and subtract.

B. $3^4 \div 3 - 5 \times 4$

$\boxed{} \div 3 - 5 \times 4$

$\boxed{} - \boxed{} = \underline{}$

D.

C. $3 \times (64 \div 4) + 2^3 - 3$

$3 \times \boxed{} + 2^3 - 3$

$3 \times \boxed{} + \boxed{} - 3$

$\boxed{} + \boxed{} - 3 = \underline{}$

Practice

▶ **Use the order of operations to simplify each expression.**

1. $12 + 9 \times 4 \div 3 =$ _____

2. $27 \div (8 - 5)^2 + 5 =$ _____

3. $(9 + 6)^2 - (7 \times 5) =$ _____

4. $4^3 \div (24 - 8) + 9 =$ _____

5. $3^4 - 2^3 \times 9 =$ _____

6. $(5^2 - 15)^2 \div 4 =$ _____

7. $49 - 56 \div 2^3 \times 3 =$ _____

8. $(2 \times 3)^2 \div (1 + 5) =$ _____

9. $2^4 \times 5 + (25 - 16) =$ _____

10. $5^3 - 12 \times (2 + 8) =$ _____

11. $(9 + 3)^2 \div 3^2 + 15 =$ _____

12. $(18 + 10) \times 2^2 \div 8 =$ _____

▶ **Determine whether each equation is true. Circle your answer.**

13. $12 + 8 \div 4 \times 3 \overset{?}{=} 18$ Yes   No

14. $9^2 - (19 + 8) \div 9 \overset{?}{=} 6$ Yes   No

15. $(4^3 \div 2) - 19 \overset{?}{=} 13$ Yes   No

16. $38 + 8 \times 3 \div 4 \overset{?}{=} 34$ Yes   No

ACTIVITY 33 Algebraic Expressions

Complete this page with your teacher.

▶ **Write each word phrase as an algebraic expression.**

Word Phrase	Expression
A. 8 less than a number	$n - 8$
B. the sum of a number and 15	_____
C. a number divided into 6 parts	_____
D. the product of 3 and a number	_____
E. 4 more than 5 times a number	$5n + 4$
F. 2 less than a number divided by 3	_____
G. 7 times the difference between a number and 15	_____
H. the difference between 9 times a number and 8	_____

▶ **Write each algebraic expression as a word phrase.**

I. $n - 6$ _____

_____

J. $4(n + 2)$ _____

_____

K. _____ _____

_____

Practice

▶ **Write each word phrase as an algebraic expression.**

1. 7 more than a number _____

2. 5 times the sum of a number and 3 _____

3. the product of 20 and a number _____

4. 12 less than a number divided by 6 _____

5. the sum of 8 times a number and 2 _____

6. the difference between 10 and a number _____

▶ **Write each algebraic expression as a word phrase.**

7. $n + 14$

_____

8. $\frac{n}{5} + 3$

_____

9. $7n$

_____

10. $2n - 9$

_____

Complete this page with your teacher.

▶ **Use the value of the variable to evaluate each expression.**

A. $n + 8$ if $n = 6$ Substitute the value for the variable and simplify.

$6 + 8 =$ ___14___

B. $4(x - 5)$ if $x = 10$ Follow the order of operations to simplify.

$4(\underline{\hspace{1.5cm}} - 5)$

$4(\underline{\hspace{1.5cm}}) = \underline{\hspace{1.5cm}}$

C. $\dfrac{d}{3} + 9$ if $d = 36$ **D.** $t^2 + 13$ if $t = 7$

E. $12s$ if $s = 9$ **F.** $8y - 24$ if $y = 4$

G. $10p - (p \times 3)$ if $p = 10$ **H.** $48 \div (5^2 - m)$ if $m = 13$

I. $\underline{\hspace{3cm}}$ if $\underline{\hspace{1.5cm}} = \underline{\hspace{1.5cm}}$

Practice

▶ **Use the value of the variable to evaluate each expression.**

1. $b - 10$ if $b = 32$ _____

2. $3g + 6$ if $g = 7$ _____

3. $5(w + 4)$ if $w = 8$ _____

4. $m^2 - 30$ if $m = 12$ _____

5. $\frac{a}{7} - 15$ if $a = 112$ _____

6. $8(c + 7)$ if $c = 2$ _____

7. $(r + 3)^2$ if $r = 5$ _____

8. $y \div 4$ if $y = 132$ _____

9. $9u - 79$ if $u = 14$ _____

10. $14h + 3$ if $h = 7$ _____

11. $150 - j^2$ if $j = 11$ _____

12. $\frac{n}{8} + 27$ if $n = 104$ _____

ACTIVITY 35 Properties

Complete this page with your teacher.

▶ **Use the Commutative Properties of Addition and Multiplication to find each missing value.**

A. $15 + 23 =$ _____ $+ 15$

B. $284 + 690 = 690 +$ _____

C. $76 \times 5 =$ _____ $\times 76$

D. $63 \times 42 = 42 \times$ _____

▶ **Use the Associative Properties of Addition and Multiplication to find each missing value.**

E. $82 + (57 + 39) = ($ _____ $+ 57) + 39$

F. $14 \times (68 \times 20) = (14 \times$ _____ $) \times 20$

G. $(91 + 75) + 83 = 91 + ($ _____ $+$ _____ $)$

H. $291 \times (370 \times 568) = ($ _____ $\times$ _____ $) \times$ _____

I. The Commutative/Associative Property of Addition tells us that

_____ = _____.

J. The Commutative/Associative Property of Multiplication tells us that

_____ = _____.

Practice

▶ **Find each missing value. Then circle the property that the sentence models.**

1. $35 \times 67 = $ _____ $\times 35$ Commutative   Associative

2. $20 + (59 + 41) = ($ _____ $+ 20) + 41$ Commutative   Associative

3. $(408 \times 113) \times 25 = (113 \times $ _____ $) \times 408$ Commutative   Associative

4. $1{,}672 + 3{,}490 = $ _____ $+ 1{,}672$ Commutative   Associative

5. $(75 + 81) + 238 = (238 + $ _____ $) + 75$ Commutative   Associative

6. $963 \times 410 = 410 \times $ _____   Commutative   Associative

7. $87 + (162 + 203) = ($ _____ $+ 87) + 162$ Commutative   Associative

8. $(50 \times 94) \times 78 = (78 \times 50) \times $ _____   Commutative   Associative

9. $282 + 167 = $ _____ $+ 282$ Commutative   Associative

10. $435 \times (781 \times 29) = (29 \times $ _____ $) \times 781$ Commutative   Associative

11. $89 \times 76 = 76 \times $ _____   Commutative   Associative

12. $(14 + 53) + 60 = (53 + 60) + $ _____   Commutative   Associative

13. $12 \times (3 \times 8) = 12 \times (8 \times $ _____ $)$ Commutative   Associative

14. $13 + 45 = 45 = $ _____   Commutative   Associative

15. $(2^2 \times 2^3) \times 3^2 = 2^2 \times ($ _____ $\times 3^2)$ Commutative   Associative

Complete this page with your teacher.

▶ **Use the Identity Property of Addition to find each sum.**

A. $56 + 0 = $ _____

B. $320 + 0 = $ _____

C. $75 + 0 = $ _____

D. $819 + 0 = $ _____

▶ **Use the Identity Property of Multiplication to find each product.**

E. $12 \times 1 = $ _____

F. $563 \times 1 = $ _____

G. $60 \times 1 = $ _____

H. $924 \times 1 = $ _____

▶ **Use the Zero Property of Multiplication to find each product.**

I. $32 \times 0 = $ _____

J. $106 \times 0 = $ _____

K. $95 \times 0 = $ _____

L. $487 \times 0 = $ _____

M. The Identity / Zero Property of Addition tells us that

_____ = _____.

N. The Identity / Zero Property of Multiplication tells us that

_____ = _____.

Practice

▶ **Write the number that makes each sentence true. Then circle the property that the sentence models.**

1. $72 \times 1 =$ _____   Zero   Identity

2. $125 + 0 =$ _____   Zero   Identity

3. $804 \times 0 =$ _____   Zero   Identity

4. $39 + 0 =$ _____   Zero   Identity

5. $671 \times 1 =$ _____   Zero   Identity

6. $253 + 0 =$ _____   Zero   Identity

7. $94 \times 17 \times 0 =$ _____   Zero   Identity

8. $468 \times 1 =$ _____   Zero   Identity

▶ **For each statement, write T for *true* or F for *false*.**

_____ 9. The Zero Property of Addition tells us that $253 + 0 = 253$.

_____ 10. The Zero Property of Multiplication tells us that $61 \times 0 = 61$.

_____ 11. The Identity Property of Multiplication tells us that $709 \times 1 = 709$

_____ 12. The Identity Property of Addition tells us that $84 + 0 = 0$.

Complete this page with your teacher.

▶ **Use the Distributive Property to rewrite each expression.**

A. $5 \times (4 + 6) = (5 \times 4) + (5 \times$ _____ $)$

B. $7(9 + 3) =$ _____ $(9) +$ _____ (3)

C. $2 \times (10 + 8) = 2($ _____ $) + 2($ _____ $)$

D. $6(12 + 15) = 6($ _____ $) +$ _____ $($ _____ $)$

E. _____ $=$ _____

▶ **Use the Distributive Property to simplify each expression.**

F. $8(5 + 9) =$

$(8 \times 5) + (8 \times 9) =$

_____ $+$ _____ $=$ _____

G. $10 \times (7 + 6) =$

_____ $\times ($ _____ $) +$

_____ $\times ($ _____ $) =$

_____ $+$ _____ $=$ _____

H. $(9 + 3) \times 2 =$

$($ _____ $\times$ _____ $) +$

$($ _____ $\times$ _____ $) =$

_____ $+$ _____ $=$ _____

I. _____ $=$

Practice

▶ **Use the Distributive Property to rewrite each expression.**

1. $3(7 + 9) =$ _____

2. $10(8 + 6) =$ _____

3. $4 \times (5 + 11) =$ _____

4. $15(3 + 8) =$ _____

▶ **Use the Distributive Property to simplify each expression. Show your work.**

5. $7(8 + 10) =$ _____

6. $8(5 + 9) =$ _____

7. $5(4 + 7) =$ _____

8. $9(10 + 6) =$ _____

9. $2(3 + 9) =$ _____

10. $6(1 + 8) =$ _____

11. $12(5 + 6) =$ _____

12. $14(7 + 5) =$ _____

13. $7(18 + 10) =$ _____

14. $20(8 + 9) =$ _____

Complete this page with your teacher.

▶ **Find the value of the variable in each equation.**

A.
$$8 \times n = 112$$
$$8 \times n \div \underline{\hspace{1.5cm}} = 112 \div \underline{\hspace{1.5cm}}$$
$$n = \underline{\hspace{1.5cm}}$$

B.
$$r - 43 = 95$$
$$r - 43 + \underline{\hspace{1.5cm}} = 95 + \underline{\hspace{1.5cm}}$$
$$r = \underline{\hspace{1.5cm}}$$

C.
$$v \div 14 = 84$$
$$v \div 14 \times \underline{\hspace{1.5cm}} = 84 \times \underline{\hspace{1.5cm}}$$
$$v = \underline{\hspace{1.5cm}}$$

D.
$$m + 25 = 61$$
$$m + 25 - \underline{\hspace{1.5cm}} = 61 - \underline{\hspace{1.5cm}}$$
$$m = \underline{\hspace{1.5cm}}$$

E.
$$\frac{s}{7} = 91$$
$$\frac{s}{7} \times \underline{\hspace{1.5cm}} = 91 \times \underline{\hspace{1.5cm}}$$
$$s = \underline{\hspace{1.5cm}}$$

F. _____

▶ **Write and solve an equation for the problem.**

G. A family bought 4 movie tickets for $26.00. How much did each ticket cost?

$$\underline{\hspace{2cm}} \bigcirc \underline{\hspace{2cm}} = \underline{\hspace{2cm}}$$

number of tickets   unknown   total cost of tickets

Practice

▶ **Find the value of the variable in each equation.**
Write the letter of the correct answer on the line.

_____ 1. $8x = 384$

A. $x = 23$

_____ 2. $x + 30 = 38$

B. $x = 288$

_____ 3. $x \div 6 = 48$

C. $x = 200$

_____ 4. $4x = 92$

D. $x = 48$

_____ 5. $x - 30 = 53$

E. $x = 83$

_____ 6. $\frac{x}{5} = 40$

F. $x = 8$

_____ 7. $605 + x = 636$

G. $x = 31$

_____ 8. $160 \div x = 10$

H. $x = 16$

▶ **Write and solve an equation for each problem.**

9. Carlos got 17 new state quarters. He now has a total of 42 state quarters. How many state quarters did he have before he got the new quarters?

_____

10. The math teacher gave 12 students a new package of graph paper to share equally. Each student received 6 sheets of paper. How many pieces of graph paper were in the package?

_____

Complete this page with your teacher.

▶ **Find the value of the variable in each equation.**

A.

$$5 + 3n = 119$$

$$5 + 3n - \underline{\hspace{1cm}} = 119 - \underline{\hspace{1cm}}$$

$$3n = \underline{\hspace{1cm}}$$

$$3n \div \underline{\hspace{1cm}} = \underline{\hspace{1cm}} \div \underline{\hspace{1cm}}$$

$$n = \underline{\hspace{1cm}}$$

- Isolate the variable.

- Use inverse operations.

- Keep the equation balanced.

B.

$$6t - 24 = 18$$

$$6t - 24 + \underline{\hspace{1cm}} = 18 + \underline{\hspace{1cm}}$$

$$6t = \underline{\hspace{1cm}}$$

$$6t \div \underline{\hspace{1cm}} = \underline{\hspace{1cm}} \div \underline{\hspace{1cm}}$$

$$t = \underline{\hspace{1cm}}$$

C.

$$7b + 184 = 485$$

$$7b + 184 - \underline{\hspace{1cm}} = 485 - \underline{\hspace{1cm}}$$

$$7b = \underline{\hspace{1cm}}$$

$$7b \div \underline{\hspace{1cm}} = \underline{\hspace{1cm}} \div \underline{\hspace{1cm}}$$

$$b = \underline{\hspace{1cm}}$$

D.

$$\frac{x}{5} + 85 = 190$$

$$\frac{x}{5} + 85 - \underline{\hspace{1cm}} = 190 - \underline{\hspace{1cm}}$$

$$\frac{x}{5} = \underline{\hspace{1cm}}$$

$$\frac{x}{5} \times \underline{\hspace{1cm}} = \underline{\hspace{1cm}} \times \underline{\hspace{1cm}}$$

$$x = \underline{\hspace{1cm}}$$

E. _____

Practice

▶ **Find the value of the variable in each equation. Show your work.**

1. $3s - 18 = 66$

2. $42 + 5y = 67$

3. $\dfrac{m}{8} + 23 = 151$

4. $9d + 240 = 582$

5. $6w - 75 = 39$

6. $38 + 4p = 254$

7. $59 + 2z = 153$

8. $\dfrac{g}{7} - 217 = 217$

9. $7h - 18 = 122$

10. $12a + 97 = 145$

Complete this page with your teacher.

▶ **Solve each inequality.**

A. $t + 10 > 14$

$t + 10 - \mathbf{10} > 14 - \mathbf{10}$

$t > 4$

Check: _____ $+ 10 > 14$

B. $-m - 15 \geq 6$

$-m - 15 +$ _____ $\geq 6 +$ _____

$-m \geq 21$

$-m \div \mathbf{-1} \geq 21 \div \mathbf{-1}$

$m \leq -21$

Check: _____ $- 15 \geq 6$

C. $y + 22 > 37$

Check: _____

D. $-5t - 10 < 80$

Check: _____

E. _____ 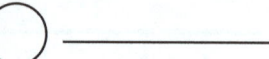 _____

Check: _____

Practice

▶ **Solve each inequality. Circle the letter of the correct answer.**

1. $x + 8 > 24$

   A. $x < 32$
   B. $x > 16$
   C. $x > 7$
   D. $x > 32$

2. $w - 53 < -15$

   A. $w < 38$
   B. $w > 38$
   C. $w < -68$
   D. $w > -68$

3. $-h - 16 \geq 70$

   A. $h \leq 54$
   B. $h \leq -86$
   C. $h \geq -54$
   D. $h \geq 86$

4. $p + 38 \leq -22$

   A. $p \leq 16$
   B. $p \geq 60$
   C. $p \geq -16$
   D. $p \leq -60$

5. $49 - r > 18$

   A. $r < 31$
   B. $r > 31$
   C. $r < -31$
   D. $r > -31$

6. $k + 92 \geq 116$

   A. $k < 208$
   B. $k > 208$
   C. $k \geq 24$
   D. $k \leq 24$

7. $-a + 87 \geq 29$

   A. $a \leq -58$
   B. $a \geq -58$
   C. $a \leq 58$
   D. $a \geq 58$

8. $n - 41 < -86$

   A. $n < -127$
   B. $n < -45$
   C. $n > 127$
   D. $n > 45$

9. $z - 13 > 91$

   A. $z < 78$
   B. $z > 78$
   C. $z < 104$
   D. $z > 104$

10. $157 - f \leq 60$

    A. $f \geq 97$
    B. $f \leq -97$
    C. $f \geq 217$
    D. $f \leq -217$

Complete this page with your teacher.

▶ **Measure and classify each angle.**

A. m∠ABC = _____°

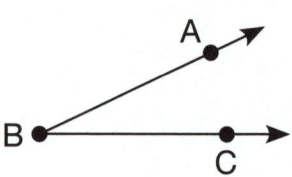

_____ angle

B. m∠XYZ = _____°

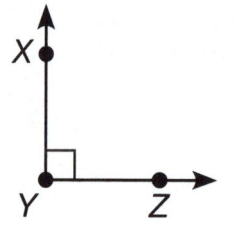

_____ angle

C. m∠DEF = _____°

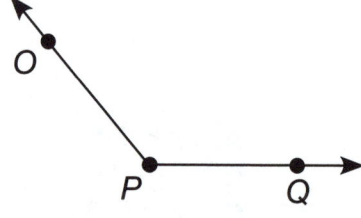

_____ angle

D. m∠OPQ = _____°

_____ angle

E. m∠RST = _____°

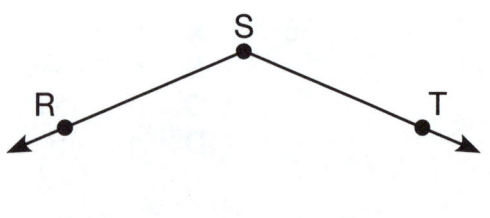

_____ angle

F. m∠JKL = _____°

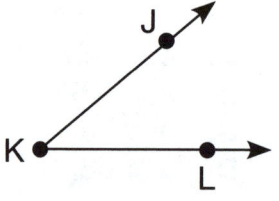

_____ angle

G. Draw an angle with a measure of _____°.
Classify the angle.

Practice

▶ **Use the figure and a protractor for Problems 1–3.**

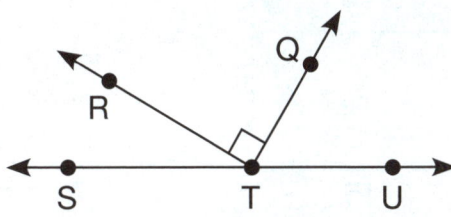

1. Name one right angle.   ∠_____

2. Name two obtuse angles.   ∠_____, ∠_____

3. Name two acute angles.   ∠_____, ∠_____

▶ **Find the measure of each angle.**

4. m∠QTU = _____°

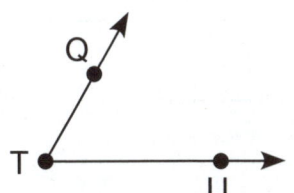

5. m∠STR = _____°

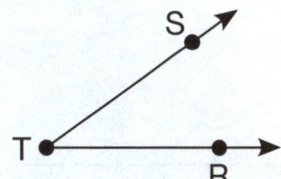

6. m∠STQ = _____°

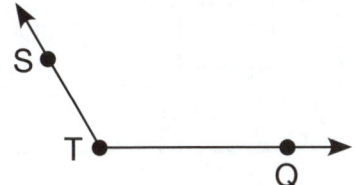

7. m∠RTU = _____°

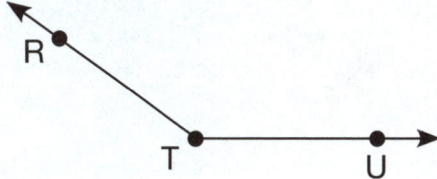

8. In the figure at the top, draw ∠STV so that it has a measure of 145°.
    What kind of angle is ∠STV?

    ∠STV is a(n) _____ angle.

Complete this page with your teacher.

▶ **Use proportions to determine if the figures are similar.**

A.

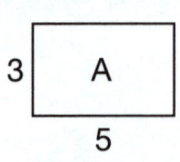

5

9 B

15

$$\frac{\text{length of } A}{\text{length of } B} \overset{?}{=} \frac{\text{width of } A}{\text{width of } B}$$

$$\frac{3 \text{ units}}{9 \text{ units}} \overset{?}{=} \frac{5 \text{ units}}{15 \text{ units}}$$

$$3 \times 15 \overset{?}{=} 5 \times 9$$

$$45 = 45$$

The sides have equal ratios, so the figures are similar.

B.

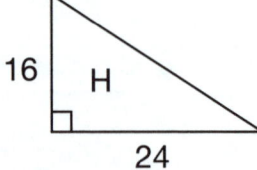

16 H

24

4 I

8

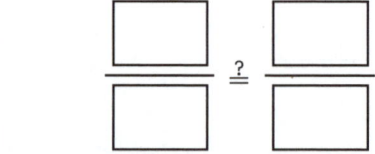

_____ × _____ $\overset{?}{=}$ _____ × _____

_____ $\overset{?}{=}$ _____

The figures are _____.

C.

_____ × _____ $\overset{?}{=}$ _____ × _____

_____ $\overset{?}{=}$ _____

The figures are _____.

Practice

▶ **Use proportions to determine if the figures are similar. Write TRUE or FALSE for each sentence.**

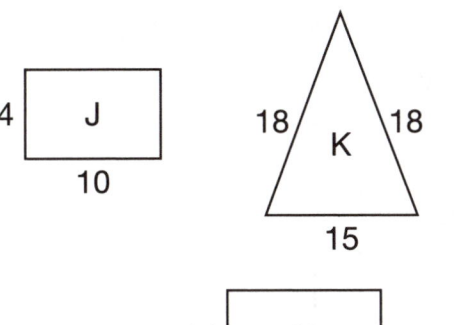 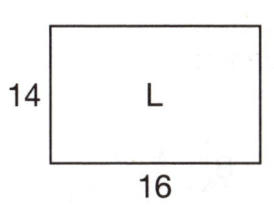 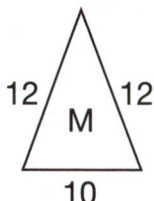

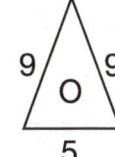

Show your work here.

_____ **1.** Figure *J* is similar to Figure *L*.

_____ **2.** Figure *P* is similar to Figure *N*.

_____ **3.** Figure *K* is similar to Figure *M*.

_____ **4.** Figure *N* is similar to Figure *L*.

_____ **5.** Figure *M* is similar to Figure *O*.

_____ **6.** Figure *J* is similar to Figure *P*.

_____ **7.** Figure *P* is similar to Figure *L*.

_____ **8.** Figure *K* is similar to Figure *O*.

Complete this page with your teacher.

A **coordinate plane** has four regions called **quadrants**.
A Roman numeral identifies each quadrant.

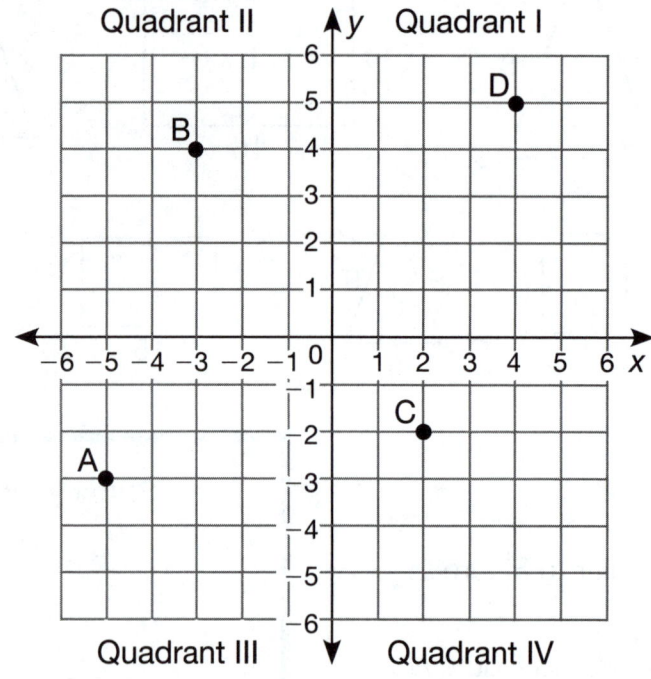

Quadrant II   Quadrant I

Quadrant III   Quadrant IV

▶ **Write the ordered pair for each point.**

A. A (_____ , _____)　　　B (_____ , _____)

C (_____ , _____)　　　D (_____ , _____)

▶ **Plot each point and identify its quadrant (Q).**

B. E (3, −4) Q _____　　　F (−4, 2)　 Q _____

G (1, 3)　 Q _____　　　H (−2, −5) Q _____

C. Ordered pair: (_____ , _____) Quadrant: ____

Practice

▶ **Write the ordered pair for each point and identify its quadrant (Q).**

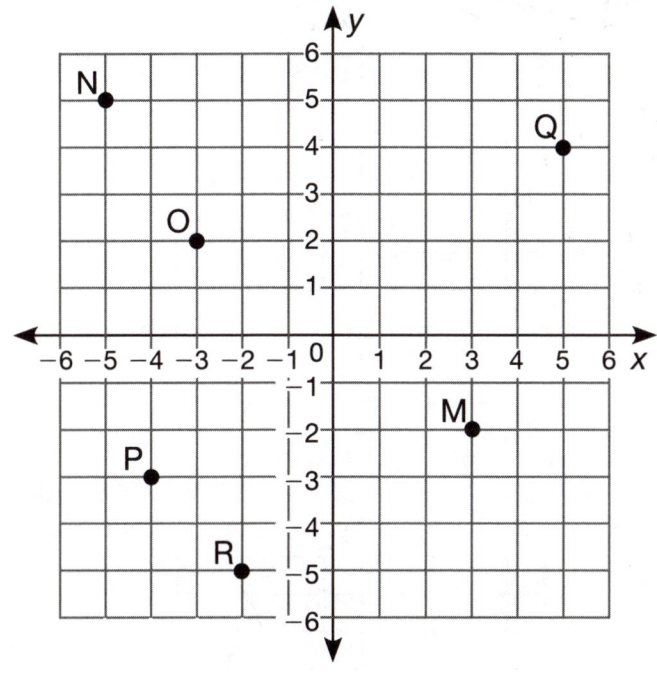

1. M (_____ , _____)   Q _____

2. N (_____ , _____)   Q _____

3. O (_____ , _____)   Q _____

4. P (_____ , _____)   Q _____

5. Q (_____ , _____)   Q _____

6. R (_____ , _____)   Q _____

▶ **Plot each point and identify its quadrant (Q).**

7. S (5, 3)   Q _____

8. T (3, −3)   Q _____

9. U (1, 1)   Q _____

10. V (−4, −5)   Q _____

11. W (−4, 3)   Q _____

12. X (4, −5)   Q _____

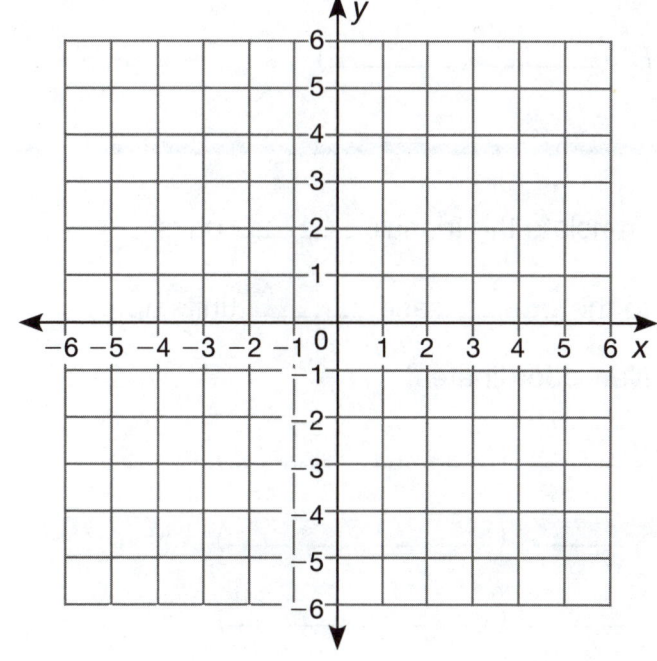

Complete this page with your teacher.

▶ **Draw each translation. Write the coordinates for the new figure.**

A. Translate the triangle 5 units to the right and 2 units up.

New coordinates:

A' (_____ , _____)

B' (_____ , _____)

C' (_____ , _____)

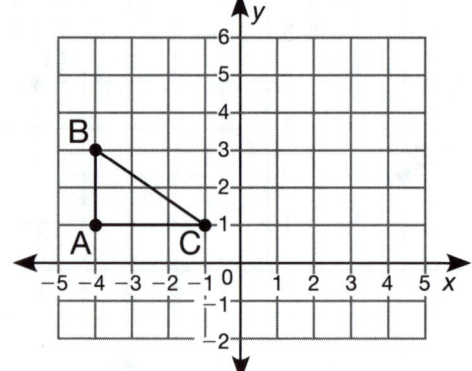

B. Translate the triangle 6 units to the left and 3 units down.

New coordinates:

F' (_____ , _____)

G' (_____ , _____)

H' (_____ , _____)

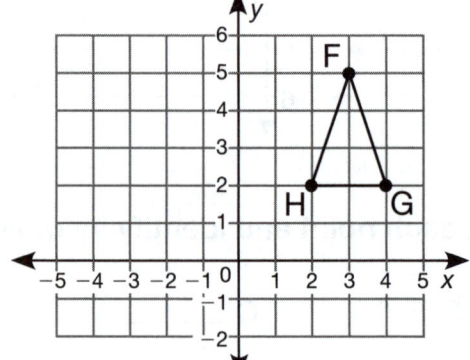

C. Translate the triangle _____ units

to the _____ and _____ units up.

New coordinates:

_____' (_____ , _____)

_____' (_____ , _____)

_____' (_____ , _____)

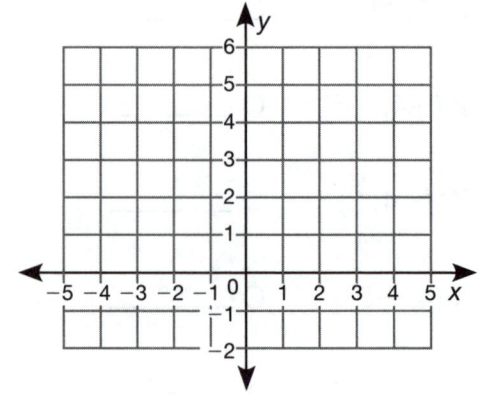

© Options Publishing. No copying permitted.

Practice

▶ **Draw each translation. Write the coordinates for the new figure.**

Translate the figure…

1. 3 units to the right and 4 units down.

New coordinates:

M′ (_____ , _____)

N′ (_____ , _____)

O′ (_____ , _____)

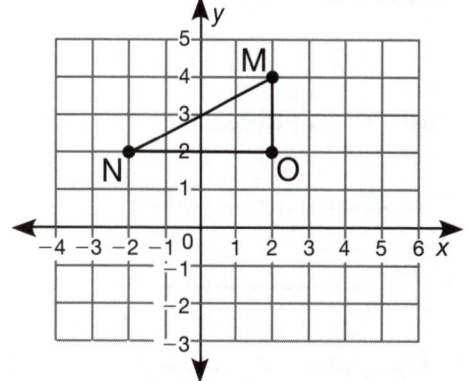

2. 5 units to the right and 3 units up.

New coordinates:

Q′ (_____ , _____)

R′ (_____ , _____)

S′ (_____ , _____)

T′ (_____ , _____)

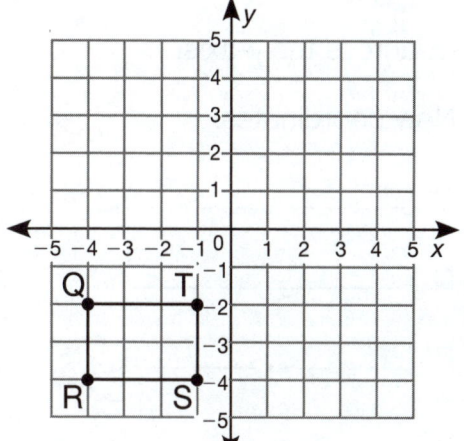

3. 2 units to the right and 3 units up.

New coordinates:

A′ (_____ , _____)

B′ (_____ , _____)

C′ (_____ , _____)

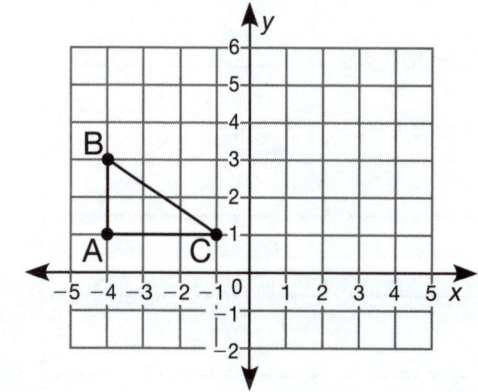

Complete this page with your teacher.

▶ **Draw each reflection. Write the coordinates for the new figure.**

Reflect the figure...

A. ... across the *x*-axis.

New coordinates:

J' (_____ , _____)

K' (_____ , _____)

L' (_____ , _____)

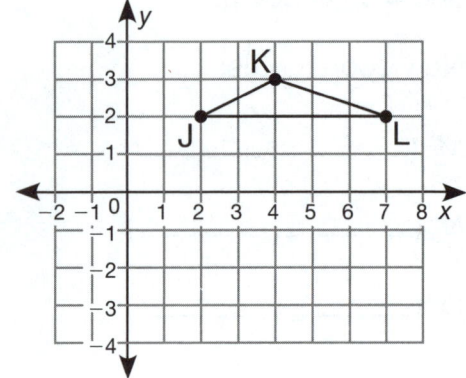

B. ... across the *y*-axis.

New coordinates:

P' (_____ , _____)

Q' (_____ , _____)

R' (_____ , _____)

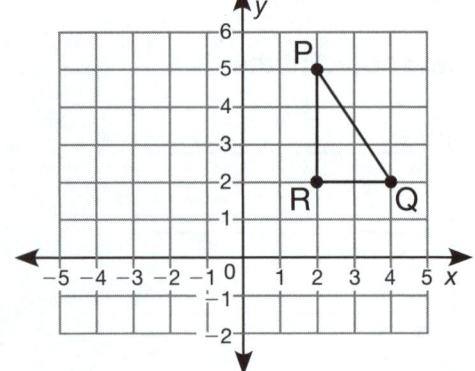

C. ... across the _____ -axis.

New coordinates:

_____' (_____ , _____)

_____' (_____ , _____)

_____' (_____ , _____)

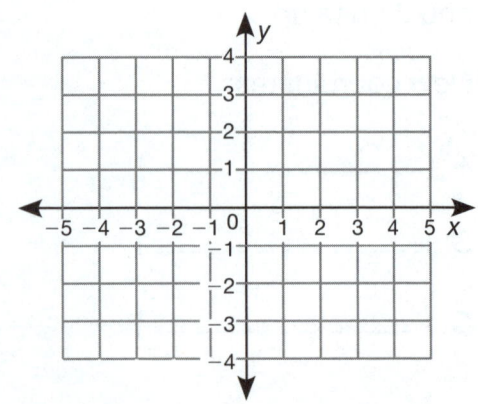

Practice

▶ **Draw each reflection. Write the coordinates for the new figure.**

Reflect the figure...

1. ... across the *x*-axis.

    New coordinates:

    B' (_____ , _____)

    C' (_____ , _____)

    D' (_____ , _____)

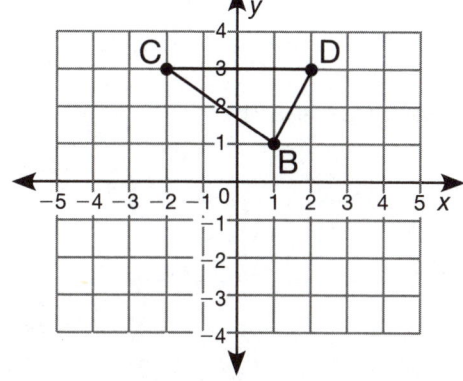

2. ... across the *y*-axis.

    New coordinates:

    G' (_____ , _____)

    H' (_____ , _____)

    I' (_____ , _____)

    J' (_____ , _____)

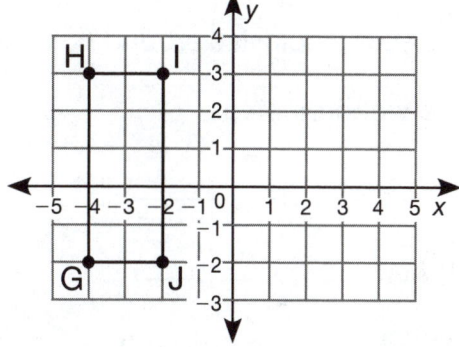

3. ... across the *x*-axis.

    New coordinates:

    X' (_____ , _____)

    Y' (_____ , _____)

    Z' (_____ , _____)

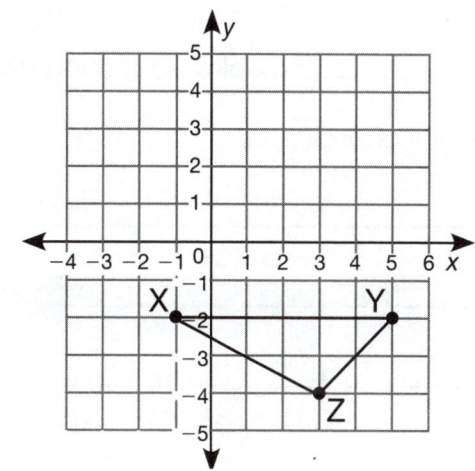

Complete this page with your teacher.

▶ **Draw each rotation. Write the coordinates for the new figure.**

Rotate the figure...

A. ... 90° clockwise around point *G*.

New coordinates:

F′ (_____, _____)

G′ (_____, _____)

H′ (_____, _____)

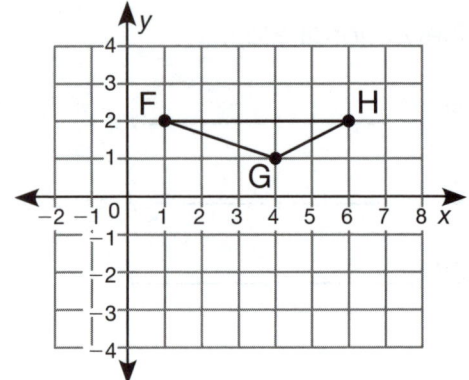

B. ... 180° clockwise around point *M*.

New coordinates:

K′ (_____, _____)

L′ (_____, _____)

M′ (_____, _____)

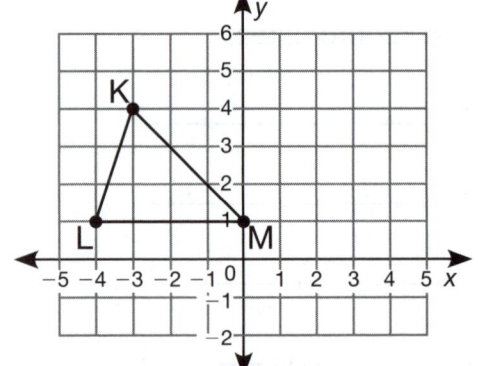

C. ... _____° clockwise around point _____.

New coordinates:

_____′ (_____, _____)

_____′ (_____, _____)

_____′ (_____, _____)

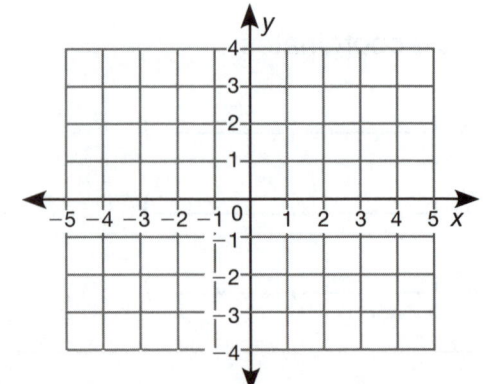

Practice

▶ **Draw each rotation. Write the coordinates for the new figure.**

Rotate the figure…

1. … 90° clockwise around point *A*.

New coordinates:

A' (_____ , _____)

B' (_____ , _____)

C' (_____ , _____)

D' (_____ , _____)

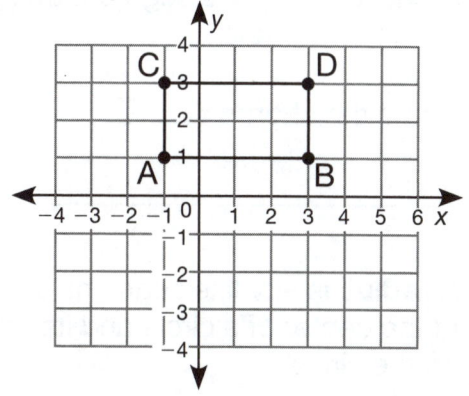

2. … 180° clockwise around point *Q*.

New coordinates:

P' (_____ , _____)

Q' (_____ , _____)

R' (_____ , _____)

S' (_____ , _____)

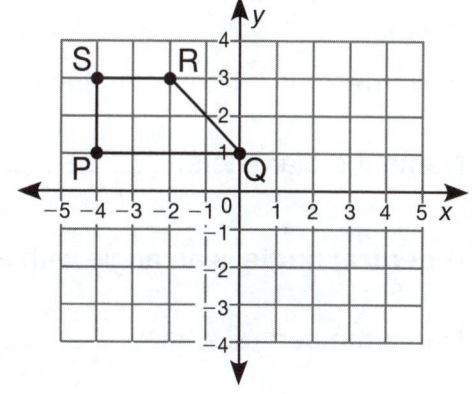

3. … 270° clockwise around point *W*.

New coordinates:

W' (_____ , _____)

X' (_____ , _____)

Y' (_____ , _____)

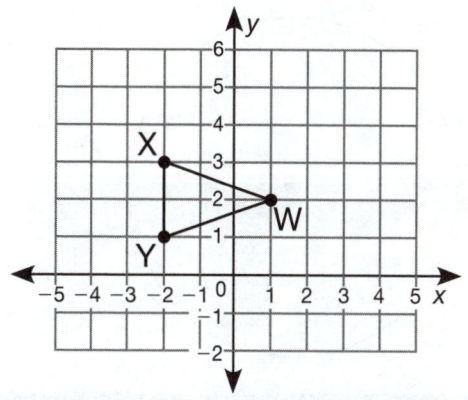

ACTIVITY (47) Parts of a Circle

Complete this page with your teacher.

▶ **Use the figure to identify parts of a circle.**

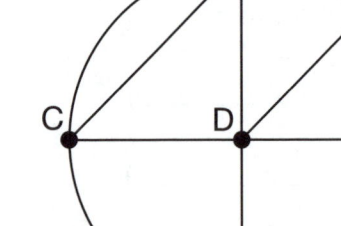

A. A **chord** is any line segment that joins two points on a circle.

Name the chords: _____

_____

B. A **radius** is any line segment with one endpoint at the center of a circle and its other endpoint on the circle.

Name the radii: _____

_____

C. A **diameter** is a chord that includes the center of the circle.

Name the diameters: _____

D. A **central angle** is an angle with a vertex that is at the center of the circle.

Name the central angles: _____

_____

E.

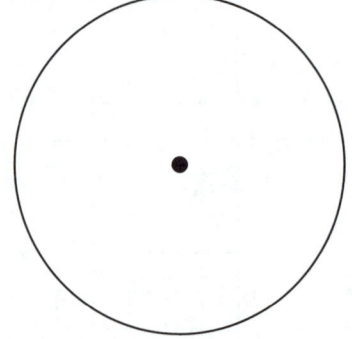

Chords: _____

Radii: _____

Diameters: _____

Central angles: _____

Practice

▶ **Use the figure to identify parts of a circle.**

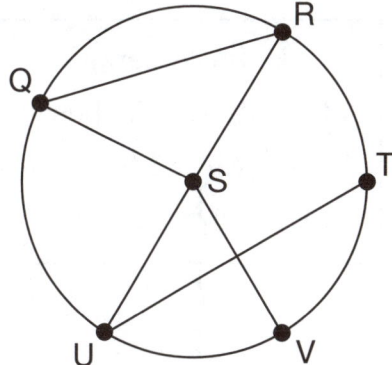

1. Chords: _____

   _____

2. Radii: _____

   _____

3. Diameter: _____

   _____

4. Central angles: _____

   _____

   _____

▶ **Circle TRUE or FALSE for each statement.**

5. TRUE   FALSE $\overline{SU}$ is a radius.

6. TRUE   FALSE $\angle TUS$ is a central angle.

7. TRUE   FALSE $\overline{UT}$ is a chord.

8. TRUE   FALSE $\overline{SR}$, $\overline{SQ}$, and $\overline{SV}$ are radii.

9. TRUE   FALSE $\angle USV$ and $\angle RQS$ are central angles.

10. TRUE   FALSE $\overline{RU}$ is a chord, a diameter, and a radius.

Complete this page with your teacher.

▶ **Convert measurement units to answer each question.**

A. How many feet are in 7 yards?

Will the number of feet be larger or smaller than 7?

_____

Which conversion fact do I need?

_____

Do I divide or multiply? _____

7 yards = _____ feet

Units of Length
1 foot (ft) = 12 inches (in.)
1 yard (yd) = 3 ft
1 mile (mi) = 5,280 ft

B. How many gallons are in 8 quarts?

Will the number of gallons be larger or smaller than 8?

_____

Which conversion fact do I need?

_____

Do I divide or multiply? _____

8 quarts = _____ gallons

Units of Capacity
1 cup (c) = 8 ounces (oz)
1 pint (pt) = 2 c
1 quart (qt) = 2 pt
1 gallon (gal) = 4 qt

C. How many _____ are in _____?

Practice

▶ **Convert measurement units to make each sentence true.**

Units of Length
1 foot (ft) = 12 inches (in.)
1 yard (yd) = 3 ft
1 mile (mi) = 5,280 ft

Units of Capacity
1 cup (c) = 8 ounces (oz)
1 pint (pt) = 2 c
1 quart (qt) = 2 pt
1 gallon (gal) = 4 qt

1. 6 mi = _____ ft

2. 5 yd = _____ ft

3. 204 in. = _____ ft

4. 26,400 ft = _____ mi

5. 22 ft = _____ in.

6. 3 mi = _____ ft

7. 36 ft = _____ yd

8. 1,760 yd = _____ mi

9. 32 oz = _____ c

10. 12 qt = _____ gal

11. 9 pt = _____ c

12. 8 c = _____ qt

13. 10 gal = _____ qt

14. 18 pt = _____ qt

15. 48 c = _____ pt

16. 16 c = _____ oz

ACTIVITY 49 Convert Metric Units

Complete this page with your teacher.

▶ **Convert measurement units to answer each question.**

Metric Prefixes	Length	Capacity
milli- = one thousandth	1 m = 1,000 mm	1 L = 1,000 mL
centi- = one hundredth	1 m = 100 cm	1,000 L = 1 kL
kilo- = one thousand	1,000 m = 1 km	

A. How many centimeters (cm) are in 9 meters (m)?

Will the number of centimeters be larger or smaller than 9? _____

Which conversion fact do I need? _____

Do I divide or multiply? _____

9 meters = _____ centimeters

B. How many liters (L) are in 12.5 kiloliters (kL)?

Will the number of liters be larger or smaller than 12.5? _____

Which conversion fact do I need? _____

Do I divide or multiply? _____

12.5 kiloliters = _____ liters

C. How many _____ are in _____?

Practice

▶ **Convert measurement units to make each sentence true.**

Length
1 m = 1,000 mm
1 m = 100 cm
1,000 m = 1 km

Capacity
1 L = 1,000 mL
1,000 L = 1 kL

1. 15 m = _____ cm

2. 450 cm = _____ m

3. 6,000 mm = _____ m

4. 70 cm = _____ mm

5. 120 m = _____ cm

6. 3.49 km = _____ mm

7. 40.08 cm = _____ m

8. 31.65 m = _____ cm

9. 3,000 mL = _____ L

10. 18 kL = _____ L

11. 9.8 L = _____ mL

12. 24.5 L = _____ mL

13. 1,650 L = _____ kL

14. 21,400 L = _____ kL

15. 6.85 L = _____ mL

16. 4.11 kL = _____ mL

▶ **Write TRUE or FALSE for each statement.**

_____ 17. 9,000 mL = 9 L

_____ 18. 7.4 km = 740 m

_____ 19. 86 m = 8,600 mm

_____ 20. 53,000 L = 53 kL

_____ 21. 620 cm = 6.2 m

_____ 22. 120 mm = 12 cm

Complete this page with your teacher.

▶ **Find the perimeter of each figure.**

A.

20 in.

14 in.   14 in.

20 in.

$P = 2l + 2w = 2(20) + 2(14)$

= _____ in.

B.

6 cm

$P = 4s = 4 \times 6$

= _____ cm

C.   18.5 m

$P =$ _____

= _____ m

D.   35 ft

59 ft

$P =$ _____

= _____ ft

E.

$P =$ _____

= _____

Practice

▶ **Find the perimeter of each figure.**

1.

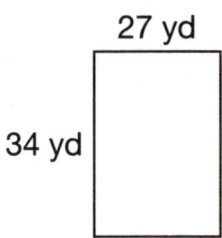

27 yd

34 yd

P = _____

= _____ yd

2.

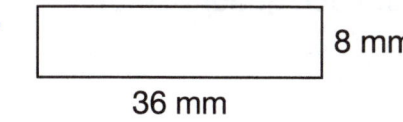

8 mm

36 mm

P = _____

= _____ mm

3.

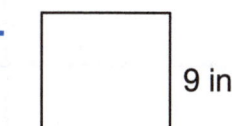

94 cm

47 cm

P = _____

= _____ cm

4.

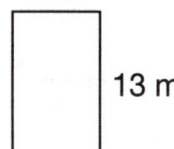

9 in.

P = _____

= _____ in.

5.

23.2 ft

P = _____

= _____ ft

6.

8 m

13 m

P = _____

= _____ m

Complete this page with your teacher.

► **Find the area of each quadrilateral.**

A. Rectangle

34 mm

20 mm

$A = l \times w$

= _____ × _____

= _____ mm²

B. Parallelogram

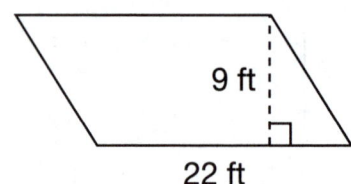

9 ft

22 ft

$A = bh$

= _____ × _____

= _____ ft²

C. Trapezoid

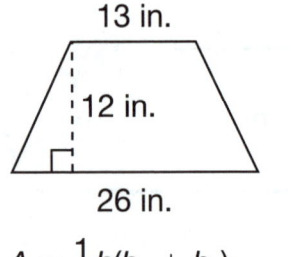

13 in.

12 in.

26 in.

$A = \frac{1}{2}h(b_1 + b_2)$

= $\frac{1}{2}$(_____)(_____ + _____)

= $\frac{1}{2}$(_____)(_____)

= _____ in.²

D. Square

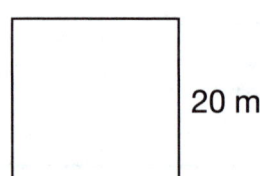

20 m

$A = s^2$

= _____²

= _____ m²

E. Dimensions: _____

$A =$ _____

Practice

▶ **Find the area of each quadrilateral.**

1.

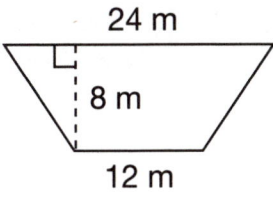

24 m

8 m

12 m

A = _____ m²

2.

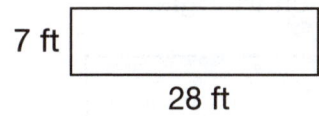

7 ft

28 ft

A = _____ ft²

3.

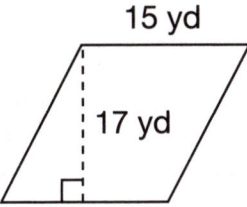

15 yd

17 yd

A = _____ yd²

4.

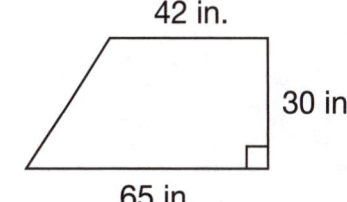

42 in.

30 in.

65 in.

A = _____ in.²

5.

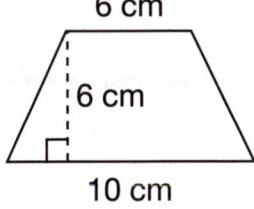

6 cm

6 cm

10 cm

A = _____ cm²

6.

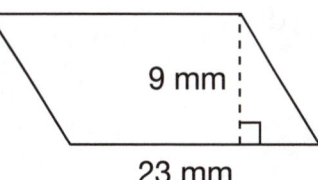

9 mm

23 mm

A = _____ mm²

Complete this page with your teacher.

▶ **Find the area of each triangle.**

A.

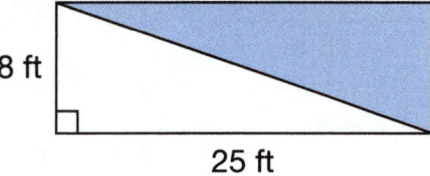

8 ft
25 ft

$A = \frac{1}{2}bh$

$= \frac{1}{2} \times 8 \times 25$

$= \underline{\hspace{1.5cm}}$ ft²

B.

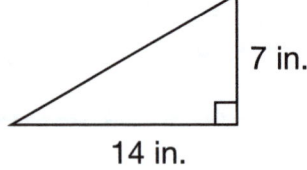

7 in.
14 in.

$A = \underline{\hspace{1.5cm}}$ in.²

C.

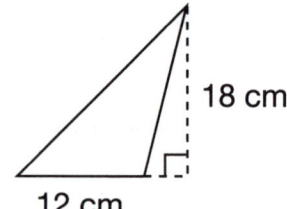

18 cm
12 cm

$A = \underline{\hspace{1.5cm}}$ cm²

D.

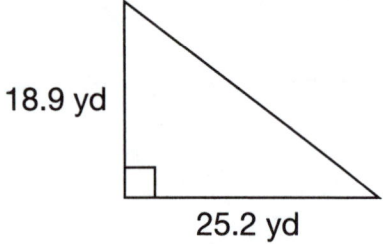

18.9 yd
25.2 yd

$A = \underline{\hspace{1.5cm}}$ yd²

E.

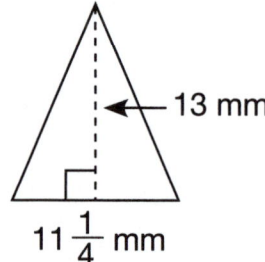

13 mm
$11\frac{1}{4}$ mm

$A = \underline{\hspace{1.5cm}}$ mm²

F.

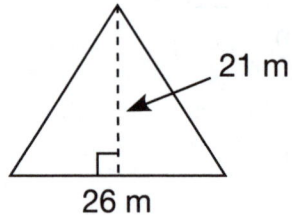

21 m
26 m

$A = \underline{\hspace{1.5cm}}$ m²

G.

$A = \underline{\hspace{3cm}}$

Practice

▶ **Find the area of each triangle.**

1.

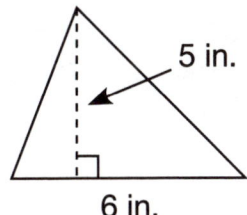

5 in.

6 in.

$A =$ _____ in.²

2.

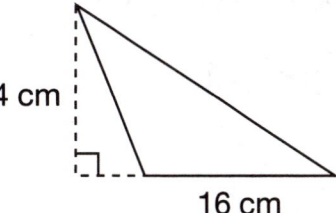

14 cm

16 cm

$A =$ _____ cm²

3.

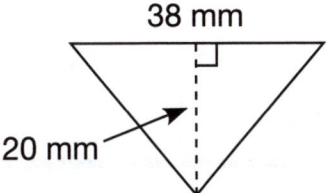

38 mm

20 mm

$A =$ _____ mm²

4.

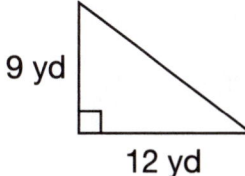

9 yd

12 yd

$A =$ _____ yd²

5.

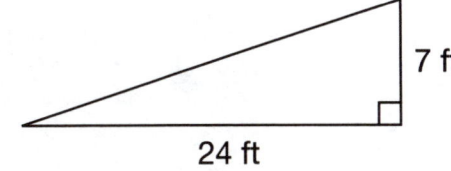

7 ft

24 ft

$A =$ _____ ft²

6.

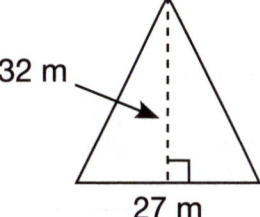

32 m

27 m

$A =$ _____ m²

7. a triangle with
$b = 17$ cm, $h = 20$ cm

$A =$ _____

8. a triangle with
$b = 2$ in., $h = 3$ in.

$A =$ _____

9. a triangle with
$b = 9$ yd, $h = 8$ yd

$A =$ _____

10. a triangle with
$b = 30$ ft, $h = 18$ ft

$A =$ _____

Complete this page with your teacher.

▶ **Find the volume of each figure.**

A.

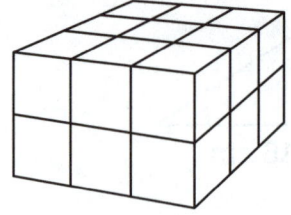

$V = lwh$

$V = 3 \times 3 \times 2$

$V = $ _____ cubic units

B.

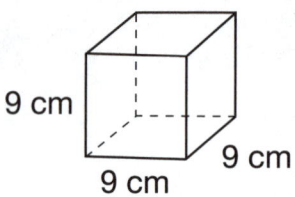

9 cm
9 cm
9 cm

$V = lwh$

$V = $ _____ × _____ × _____

$V = $ _____ cm³

C.

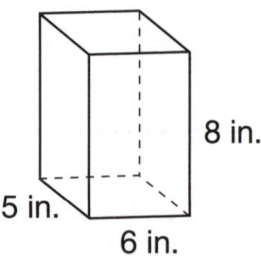

8 in.
5 in.
6 in.

$V = lwh$

$V = $ _____ × _____ × _____

$V = $ _____ in.³

D.

$V = $ _____

▶ **Solve.**

E. A box has a volume of 1,143.8 cubic mm. If its length is 21.5 mm, and its width is 13.3 mm, what is the height of the box?

_____

Practice

▶ **Find the volume of each rectangular prism.**

1.

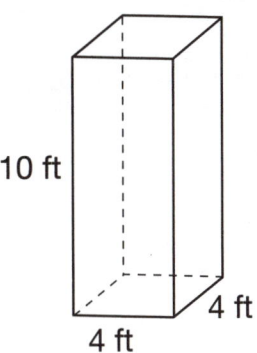

10 ft

4 ft

4 ft

$V =$ _____ ft³

2.

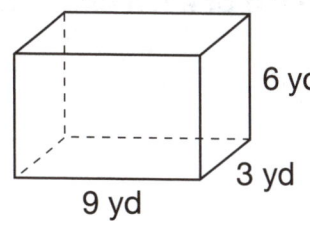

6 yd

3 yd

9 yd

$V =$ _____ yd³

3.

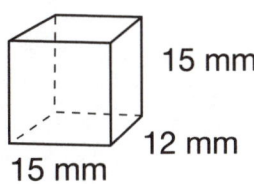

15 mm

12 mm

15 mm

$V =$ _____ mm³

4.

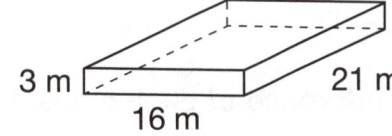

3 m

21 m

16 m

$V =$ _____ m³

▶ **Solve each problem.**

5. A student's locker measures 1.5 feet wide, 2 feet deep, and 5 feet tall.
What is the volume of the locker?

_____

6. A box measures 4 yards wide, 2.5 yards long, and 3 yards tall.
What is the box's volume?

_____

7. A crate has a volume of 504 cm³. If its length is 7 cm, and its width is 8 cm,
what is the height of the crate?

_____

Complete this page with your teacher.

▶ **Find the circumference of each circle. Use $\frac{22}{7}$ for π.**

A.

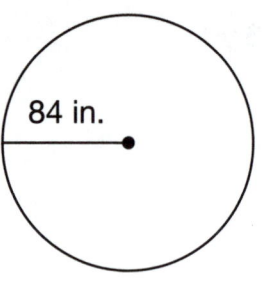

84 in.

$C = \pi d$

$C \approx \frac{22}{7} \times (2 \times 84 \text{ in.})$

$C \approx$ _____ in.

B.

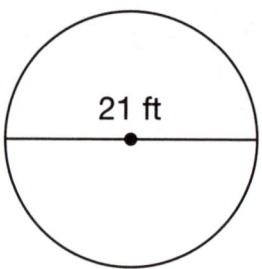

21 ft

$C = \pi d$

$C \approx \frac{22}{7} \times$ _____ ft

$C \approx$ _____ ft

▶ **Find the circumference of each circle. Use 3.14 for π.**

C.

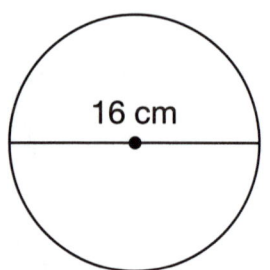

16 cm

$C = \pi d$

$C \approx 3.14 \times$ _____ cm

$C \approx$ _____ cm

D.

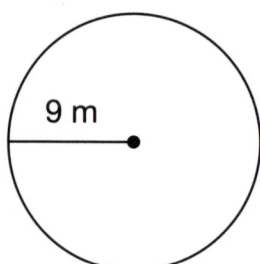

9 m

$C = \pi d$

$C \approx 3.14 \times (2 \times$ _____ m)

$C \approx$ _____ m

E.

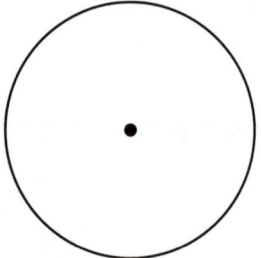

$C \approx$ _____

Practice

▶ **Find the circumference of each circle. Use 3.14 for π.**

1.

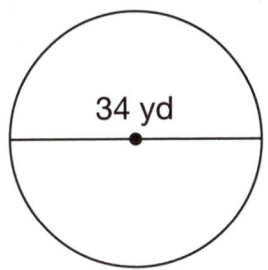

34 yd

$C \approx$ _____ yd

2.

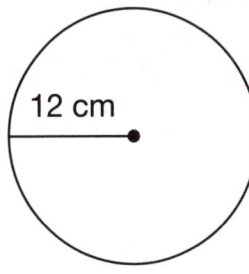

12 cm

$C \approx$ _____ cm

3.

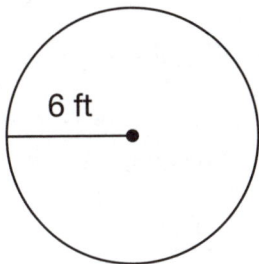

6 ft

$C \approx$ _____ ft

4.

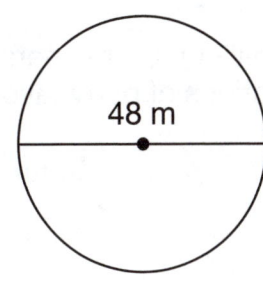

48 m

$C \approx$ _____ m

5. Find the circumference of a circle with a diameter of 91 inches.
Use $\frac{22}{7}$ for π.

_____

6. Find the circumference of a circle with a diameter of 35 feet.
Use $\frac{22}{7}$ for π.

_____

7. Find the diameter of a circle with a circumference of 78.5 cm.
Use 3.14 for π.

_____

Complete this page with your teacher.

▶ **List all of the possible outcomes.**

A. How many ways can Max, Nate, and Vick line up for a picture?

Let M = Max, N = Nate, and V = Vick.

__MNV__   _____   _____

_____   _____   _____

There are _____ possible ways to line up.

B. A party store sells balloons, streamers, and confetti in red, yellow, and green. How many combinations of party favors are there?

Favor	Color	Results
	Red →	BR
Balloons	Yellow →	BY
	Green →	BG

There are _____ possible combinations.

Practice

▶ **List all of the possible outcomes.**

1. Four students take turns answering questions. If they rotate their order, how many different combinations are possible?

2. A summer camp offers classes and outdoor activities. The classes are painting, music, dance, or theater. Outdoor activities include archery, hiking, cycling, or rock climbing. If campers take one class and one activity at a time, how many different combinations are possible?

3. A deli's lunch special offers egg or tuna salad sandwiches with fruit or yogurt and slaw or salad. How many different lunch combinations are possible?

4. Find the total number of uniform combinations if the choices include black, khaki, or denim shorts with a white, red, or blue T-shirt and sandals or tennis shoes.

Complete this page with your teacher.

▶ **Find each probability and its complement.**

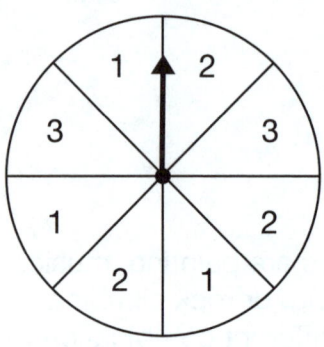

What is the probability of...

A. spinning a 2?

$$P(2) = \frac{\text{number of 2s}}{\text{number of sections}} = \underline{\hspace{2cm}}$$

B. **not** spinning a 2?

$$P(\text{not } 2) = 1 - \frac{\text{number of 2s}}{\text{number of sections}} = \underline{\hspace{2cm}}$$

What is the probability of...

C. spinning a square?

$$P(\text{square}) = \frac{\text{number of squares}}{\text{number of sections}} = \underline{\hspace{2cm}}$$

D. **not** spinning a square?

$$P(\text{not square}) = 1 - \frac{\text{number of squares}}{\text{number of sections}} = \underline{\hspace{2cm}}$$

E.

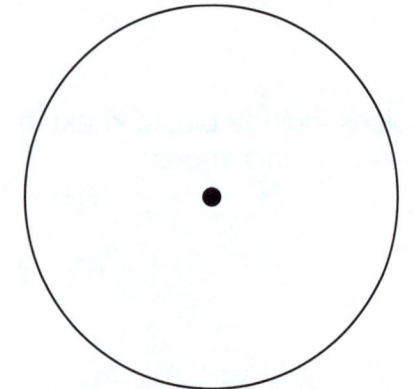

$$P(\underline{\hspace{2cm}}) = \underline{\hspace{2cm}}$$

$$P(\text{not } \underline{\hspace{2cm}}) = \underline{\hspace{2cm}}$$

Practice

▶ **Find each probability.**

1. Using the spinner on the right, find the probability of landing on an even number and its complement.

   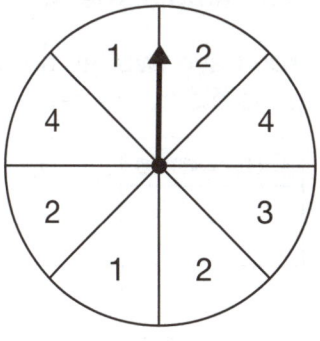

   P(even number) = _____

   P(not even number) = _____

2. One number cube is labeled 1 through 6. Find the probability of rolling a 5 and its complement.

   P(5) = _____   *P*(not 5) = _____

3. A bag contains 15 red, 16 blue, and 17 yellow marbles. Find the probability of randomly selecting a blue marble and its complement.

   P(blue) = _____   *P*(not blue) = _____

4. A hat contains 10 orange tickets, 12 green tickets, 14 purple tickets, and 8 white tickets. Find the probability of randomly selecting a ticket that is **not** white.

   P(not white ticket) = _____

5. A deck of 35 number cards is labeled 1–35. Find the probability of each of these events.

   P(18) = _____   *P*(odd number) = _____

   P(less than 6) = _____   *P*(not less than 6) = _____

Complete this page with your teacher.

► **Make a sample space to show all of the possible outcomes.**

A. You have two fair number cubes labeled 1 through 6. Find all of the possible outcomes from one roll of each cube.

Outcomes	1	2	3	4	5	6
1	1, 1	1, 2	1, 3	1, 4	1, 5	1, 6
2	2, 1					
3						
4						
5						
6						

There are _____ possible outcomes.

B. Three fair coins are tossed. Find all of the possible outcomes from the coin toss.

Coin 1	Coin 2	Coin 3	Results
		Heads	HHH
	Heads	Tails	HHT
Heads		Heads	_____
	Tails	Tails	_____
		_____	_____
_____	_____	_____	_____
	_____	_____	_____
	_____	_____	_____

There are _____ possible outcomes.

Practice

▶ **Make a sample space to show all of the possible outcomes.**

1. A board game uses one number cube labeled 1–6 and one fair coin. If a player has to toss the coin and roll the number cube per turn, how many outcomes are possible?

2. You have two shuffled decks of 7 number cards, each labeled 1–7. If you randomly draw one card from each deck per turn, how many different outcomes are possible?

3. A game asks you to toss one fair coin and randomly draw one number card from a shuffled deck of cards labeled 1–9. If a player has to toss the coin and draw one card per turn, how many outcomes are possible?

4. A game uses two fair coins and one number cube labeled 1–6. If a player has to toss both coins and roll the number cube per turn, how many outcomes are possible?

Complete this page with your teacher.

▶ **Find each probability.**

A. You have two shuffled decks of 4 number cards labeled 1–4. What is the probability of randomly drawing a 2 from the first deck and an odd number from the second?

Second Deck

Cards	1	2	3	4
1	1, 1	1, 2	1, 3	1, 4
2	2, 1			
3				
4				

(First Deck label on left side)

$P(2, \text{odd number}) = \frac{1}{4} \times \frac{1}{2} = \underline{\hspace{1.5cm}}$

B. Find the probability of randomly drawing a 3 from the first deck and a 4 from the second deck.

$P(3, 4) = \underline{\hspace{1.5cm}} \times \underline{\hspace{1.5cm}} = \underline{\hspace{1.5cm}}$

C. Find the probability of randomly drawing an even number from the first deck and **not** a 1 from the second deck.

$P(\text{even number, not 1}) = \underline{\hspace{1.5cm}} \times \underline{\hspace{1.5cm}} = \underline{\hspace{1.5cm}}$

D. _____

_____

$P(\underline{\hspace{1cm}}, \underline{\hspace{1cm}}) = \underline{\hspace{1.5cm}} \times \underline{\hspace{1.5cm}} = \underline{\hspace{1.5cm}}$

Practice

▶ **Find each probability.**

1. You have two shuffled decks of 8 number cards, each labeled 1–8. Find the probability of randomly drawing a 7 from the first deck and an 8 from the second deck.

   $P(7, 8) =$ _____ × _____ = _____

2. You have two fair number cubes labeled 1–6. Find the probability of rolling a 2 on the first cube and not a 2 on the second cube.

   $P(2, \text{not } 2) =$ _____ × _____ = _____

3. A game asks players to toss one fair coin and randomly draw one number card from a shuffled deck of cards labeled 1–15. Find the probability of landing on tails and drawing a card greater than 6.

   $P(\text{tails, greater than 6}) =$ _____ × _____ = _____ = _____

4. You toss two fair coins and roll one number cube labeled 1–6. Find the probability of both coins landing on heads and rolling an odd number.

   $P(\text{heads, heads, odd number}) =$

   _____ × _____ × _____ = _____ = _____

5. A game uses one fair coin, one number cube labeled 1–6, and a shuffled deck of number cards labeled 1–25. Find the probability of a player landing on tails, rolling a 3, and drawing a card greater than 5.

   $P(\text{tails, 3, greater than 5}) =$

   _____ × _____ × _____ = _____ = _____

Complete this page with your teacher.

▶ **Use the data to create a stem-and-leaf plot. Then answer the questions.**

A. The quiz scores for students in a math class are listed below.
Order the data from least to greatest.

72, 85, 94, 83, 77, 92, 86, 92, 79, 83, 88, 90, 84, 95

_____

B. Choose the stems. Write the stems on the left side of the plot.

Math Quiz Scores

Key: _____

C. Write the leaves on each stem, from least to greatest.

D. Include a key to explain how to read the stems and leaves.

E. _____

Practice

▶ **Use the data to create stem-and-leaf plots. Then answer the questions.**

These are the heights (in inches) of students in Room 106:
54, 62, 70, 65, 59, 61, 63, 60, 71, 68, 61, 58, 62, 67, 70, 64

1. Create a stem-and-leaf plot of the students' heights. Include a title and a key.

Key: _____

2. How many students are 62 inches tall? _____

3. How many students are greater than 64 inches tall? _____

4. How many students are less than 70 inches tall? _____

These are the number of tickets each student sold:
19, 43, 38, 22, 30, 18, 45, 32, 27, 41, 35, 29, 40, 25

5. Create a stem-and-leaf plot of the number of tickets sold per student. Include a title and a key.

Key: _____

6. How many students sold 10–29 tickets? _____

7. How many students sold 30–49 tickets? _____

8. Did more students sell above 30 tickets or below 30 tickets? _____

Complete this page with your teacher.

▶ **Use the double bar graph to answer the questions.**

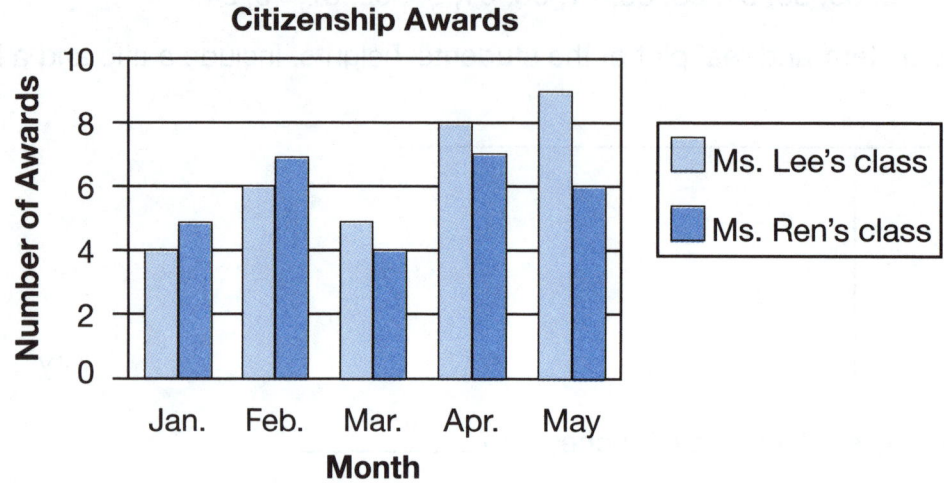

A. How many students in Ms. Lee's class earned Citizenship Awards in April?

_____

B. How many total students earned Citizenship Awards in February and March?

_____

C. How many of Ms. Ren's students earned Citizenship Awards during these five months?

_____

D. _____

_____

_____

_____

Practice

▶ **Use the double bar graph to answer the questions.**

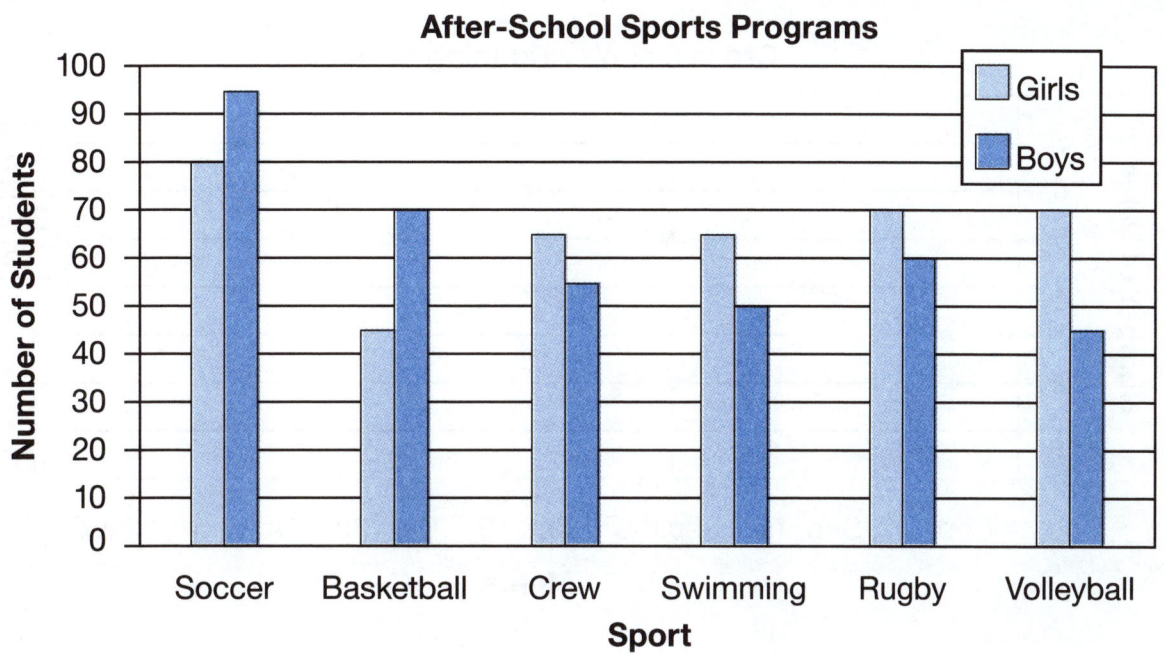

After-School Sports Programs

1. How many female students play rugby? _____

2. How many male students play soccer? _____

3. How many students in all play basketball? _____

4. Which sport has more total students, rugby or volleyball? _____

5. How many more male students are on the rugby team than the swim team?

   _____

6. How many female students are on the crew and the basketball teams?

   _____

Complete this page with your teacher.

▶ **Use the double line graph to answer the questions.**

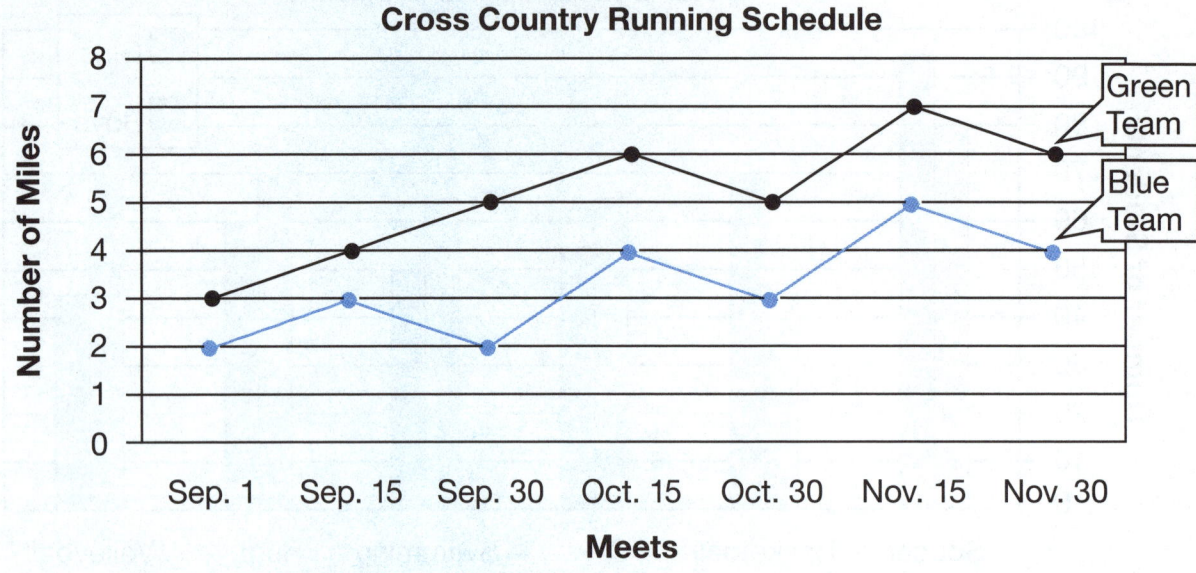

A. How many total miles did the Green and Blue Teams run during the October 30 meet?

_____

B. How many more miles did the Green Team run than the Blue Team at the

September 30 meet? _____

C. Predict how many miles the Blue Team will run at a December 15 meet.

_____

D. _____

_____

_____

Practice

▶ **Use the double line graph to answer the questions.**

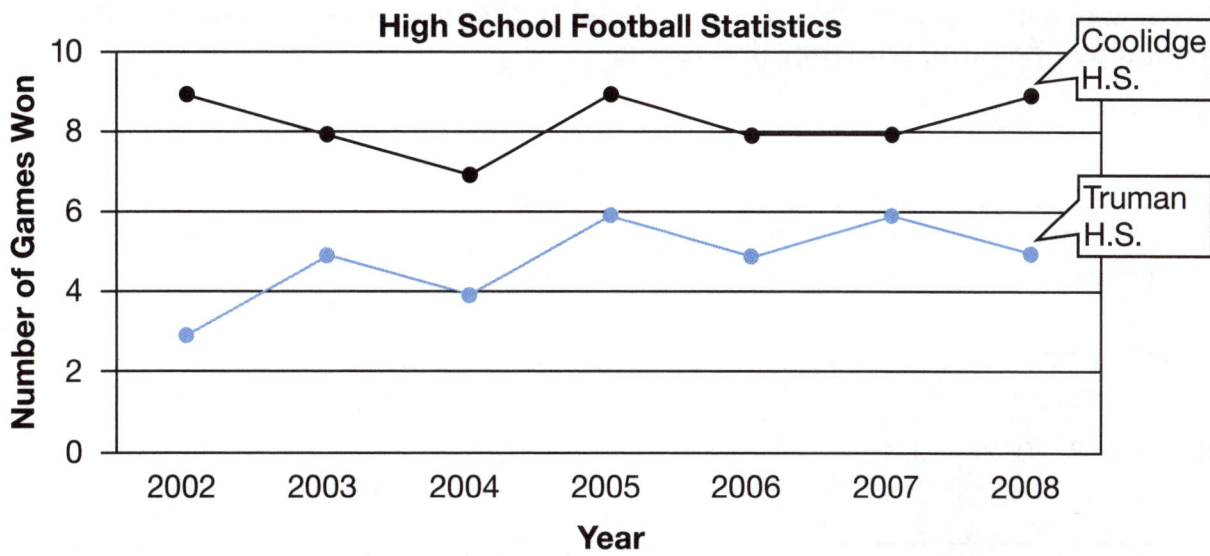

1. How many total games did the two high school teams win in the 2005 season?

   _____

2. How many more games did Coolidge H.S. win in 2002 than Truman H.S.?

   _____

3. In which year did Truman H.S. win the fewest amount of games?

   _____

4. In which two years did Truman H.S. win exactly 6 games? _____

5. Based on their pattern, predict how many games Coolidge H.S. will win in the 2009 season.

   _____

6. Based on their pattern, predict how many games Truman H.S. will win in the 2009 season.

   _____

Complete this page with your teacher.

▶ **Order the values in each data set from least to greatest. Then find each set's measures of central tendency and range.**

A. 35, 79, 56, 65, 87, 56

_____ , _____ , _____ , _____ , _____ , _____

Mean: _____   Mode: _____

Median: _____   Range: _____

B. 24, 18, 30, 24, 30, 12

_____ , _____ , _____ , _____ , _____ , _____

Mean: _____   Mode: _____

Median: _____   Range: _____

C. 114, 406, 379, 105, 257, 383

_____ , _____ , _____ , _____ , _____ , _____

Mean: _____   Mode: _____

Median: _____   Range: _____

D. _____

_____

Mean: _____   Mode: _____

Median: _____   Range: _____

Practice

▶ **Use this set of data for Problems 1–2.**

Distance Run in Miles: 10, 5, 9, 16, 7, 8, 9, 11, 6

1. Find the set's measures of central tendency and range.

   Mean: _____   Mode: _____

   Median: _____   Range: _____

2. Explain how the mean is affected if the greatest distance is dropped.

   _____

   _____

▶ **Use this set of data for Problems 3–4.**

Scores: 86, 97, 58, 81, 97, 62, 93, 90

3. Find the set's measures of central tendency and range.

   Mean: _____   Mode: _____

   Median: _____   Range: _____

4. How do the two lowest scores affect the mean?

   _____

   _____

5. What does a large range indicate?

   _____

   _____

Common Figures and Formulas

AREA (A)

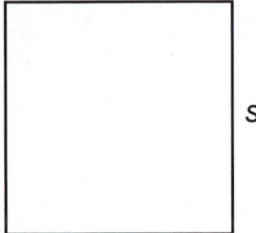

Square

$A = s^2 = s \times s$

where s = side length

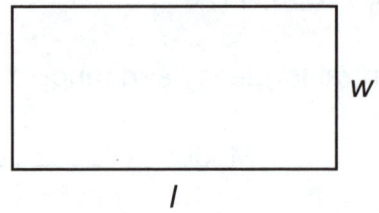

Rectangle

$A = lw = l \times w$

where l = *length*

w = *width*

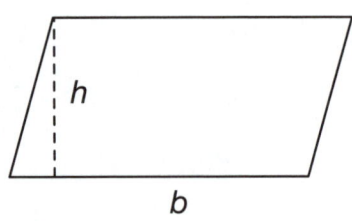

Parallelogram

$A = bh$

where b = base

h = height

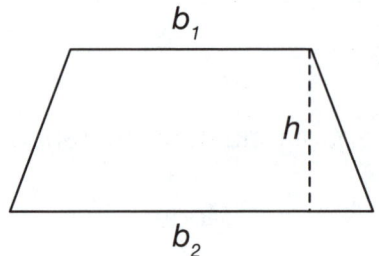

Trapezoid

$A = \frac{1}{2}h(b_1 + b_2)$

where b_1 = top base

b_2 = bottom base

h = height

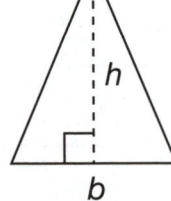

Triangle

$A = \frac{1}{2}bh$

where b = base

h = height

VOLUME (V)

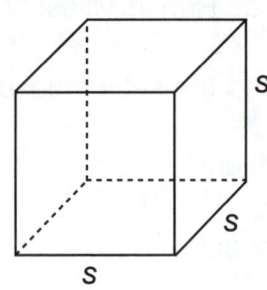

Cube

$V = s^3 = s \times s \times s$

where s = side length

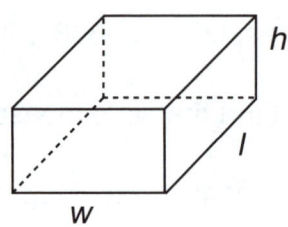

Rectangular Prism

$V = lwh = l \times w \times h$

where l = length

w = width

h = height